The Fighting America Flying Boats of WWI | Volume 1

A Centennial Perspective on Great War Seaplanes

Colin A. Owers

Great War Aviation Centennial Series #22

To G. Stuart Leslie, gentleman and friend. So much of WWI aviation history has been saved due to the efforts of Stuart in copying the photograph collections of WWI aviators.

Acknowledgements

To friends old and new. Amy Brennan; Carl Bobrow; the late J.M. Bruce; Mark Evans; Simon Frost; Frits Gerderson; Joe Gertler; Hill Goodspeed; the late Peter M. Grosz; Simon Gunn; Dan Hagedon, Adrian Hellwig; Jack Herris; Colin and Barbara Huston; Sir Peter Jackson; R. Kinter; Peter London; Bob Pearson; Bryan Ribbans; Ray Rimell; Noel Shirley; the later Ian R. Stair, Bill Toohey; the late Harry Woodman; the staffs of the RAF Museum, Hendon; the UK and USA National Archives; the Emil Buehler Library, US National Naval Aviation Museum, Pensacola; the Museum of Flight, Seattle; the Curtiss Museum, Hammondsport; and of the Aviation Heritage Trust of New Zealand; and special thanks to my wife Julie who copied many documents that were used in the compilation of this book.

Thanks to Jim Miller for colorizing the photograph used for the cover and Hubert Cance for the H-16 cutaway drawing.

Color aircraft profiles © Bob Pearson. While every care is taken, some colors are speculative based on known practices at the time. Purchase Bob's CD of WWI aircraft profiles for $50 US/Canadian, 40 €, or £30, airmail postage included, via Paypal to Bob at: **bpearson@kaien.net**

ISBN: **978-1-935881-44-5**

Design and layout: Jack Herris
Cover design: Aaron Weaver
Digital photo editing: Aaron Weaver & Jack Herris

Aeronaut Books

www.aeronautbooks.com

Table of Contents

Above: A Felixstowe *Large America* on the point of alighting. The ability of manufacturers to build the Porte type hull without being boat builders saw the number of these *Large America* flying boats increase in the last year of the War, but there were never enough to fulfil the demands made of them.

Foreword

The Royal Naval Air Service (RNAS) utilised floatplanes in the main in the years up to 1914. Flying boats were few in comparison and included the Sopwith Bat Boat. Considering that Great Britain was a maritime power, it took a long time for this nation to develop a true fighting flying boat. As will be related this came about through the offer of a prize of £10,000 by the Daily Mail for the first flight across the Atlantic. The challenge was taken up by the American Glen Curtiss who had pioneered flying boats. In the years from 1914-1918 there was a continuing exchange of ideas between the RNAS and the Curtiss Co in the USA. The Curtiss Company, in collaboration with the USN, would go on to design the NC flying boats that would cross the Atlantic in 1919. The USN was well aware of the experimental work carried on at the RNAS Felixstowe naval air station under John Porte but appear to have not had access until the US declared war on Germany. From then the USN's design branch was in contact with Porte and considered that he was the expert on large flying boats.

There were two problems facing the Royal Navy on the outbreak of war. The U-Boat threat to merchant shipping, a threat that would continue for the whole period of the war, and secondly, the Germans had Zeppelins to scout for the High Seas Fleet. It was thought that the use of the airships would give the Germans the advantage of knowing the battle formation and position of British ships if an engagement took place. This latter threat was to be overcome by the introduction of anti-Zeppelin ammunition that could be used by the standard machine guns of the day. In the battle against the Zeppelins the flying boats of the RNAS would play their part. Flying boats were the only aircraft capable of carrying out the long range reconnaissance of the German fleet stations and of the continually changing mine fields, both Allied and German.

The necessity for the large flying boat was appreciated very early by the RNAS on the outbreak of war. The existing machines were incapable of carrying ordnance that could destroy a U-Boat. As a maritime nation Great Britain required free access for its shipping and the U-Boat posed a real threat to the shipping operations that kept the country fed and supplied with the products and personnel to wage war.

On the outbreak of the war the only operational flying boat in the RNAS was the 1913 Sopwith Bat Boat amphibian. This machine had a 21 ft hull and the tail was carried on booms. The RNAS had acquired its first flying boat in 1912, an 80-hp Gnome powered Donnet-Lévêque. The service utilised floatplanes, most being conversions of existing landplanes such as the Farman series. It was to be an American who was to provide the RNAS with the weapon it required to meet the threat of the U-boat and the Zeppelin.

The American was the aviation pioneer Glen Curtiss. He was assisted by John Porte, late RN, and by Murray Sueter who had the vision to see that the large flying boat would be required for operations against the submarine menace. Sueter supported Porte at Felixstowe where the early Curtiss flying boats were used for experiments that were to lead to the successful Felixstowe series of flying boats.

It was found that the British small non-rigid airships were too slow and vulnerable for reconnaissance in the North Sea and the fighting flying boats took over this role as well as patrolling on convoy and anti-submarine patrols. Porte developed a successful flying boat that could operate from sheltered waters and that could be produced in bulk under the conditions prevailing during the war. His design was to be criticised after the war and, while it had its faults, its operational record proved it the most successful flying boat of the war. In additional to the enormous amount of work in redesigning the Curtiss flying boats, Porte designed and built the three-engined Porte Baby. While this boat eventually proved unsuccessful as a fighting boat, it continued to carry out patrols where it would not suffer interference from German aircraft. At the war's end, Porte had the huge five-engined Felixstowe Fury ready for testing. His object was to develop a boat that would be able to operate in the North Sea.

As with so much of early aviation history, the story of the development of these flying boats is confused or lacking. The operational history is well covered but the background as to the everyday problems and decisions made is overcome them is not well documented. Even the identity of the person/persons who designed these flying boats is open to question and was to be the subject of civil action in the courts after the end of the war. Court action was brought against John Porte while the war raged. Porte was accused of accepting commissions from the Curtiss Company while a serving officer and the ensuing trial was carefully covered in the aviation press of the time.

I hope that his volume will enlighten those who have never seen a majestic flying boat in flight and pay honour to the men who flew them in two World Wars.

Colin A Owers
2016

Introduction

Above: View from the front of a *Large America* flying boat. The mounting of twin Lewis guns was not common, one being the usual armament. The ammunition drums are not mounted on these guns.

The start of WWI saw European nations fall into a war that was going to be over by Christmas. The stark reality of a modern industrial war was to cause the war to linger for almost five years and involve the Middle East and eventually the United States, truly the first World War.

What is not generally known is that the United States provided aircraft to Britain from the offset of the war. Aircraft were held not to be war contraband and aircraft manufacturers in the US sold their products to the UK, most being purchased by the Royal Naval Air Service (RNAS). Glenn Curtiss was the most prominent manufacturer of flying boats in the world at the time the war broke out, and as will be related, had constructed the largest flying boat in the world with the idea of making the first flight across the Atlantic. The involvement of John Porte in this venture gave Curtiss a spokesman in the RNAS that led to a remarkable cross fertilisation of ideas that saw the Curtiss and Felixstowe flying boats become the best flying boats of their era and beyond. The full story of the association of Porte and Curtiss remains to be told. In the following chapters I have tried to tell the story from the participants view where possible. Many areas require further research

The work of the anti-submarine patrols was and is long, lonely and not often broken by periods of intense activity when a U-Boat was sighted. The *Large America* flying boats were also used for anti-Zeppelin weapons, and for long range reconnaissance flights that became more important as the war progressed. They came into combat with their German opponents who used fighting floatplanes rather than flying boats. The participation of Canadians in the air services, in one of those remarkable occurrences, saw many join the RNAS and a large proportion end up flying the big flying boats and their exploits are recorded herein.

The US entry into the war saw the anti-submarine role of the USN have the highest priority and the air service of the USN grew in a short time from a training organisation to providing Naval Air Stations in France and Britain where they flew anti-submarine patrols; and in Italy where they flew bomber and fighter flying boats of Italian design. In France the US developed Curtiss HS-1L flying boat was used, while those stations in Britain used the Curtiss H-16 *Large America* boats.

Right: Post-war scene of a Felixstowe boat flying over the ships in Harwich Harbour.

Above: Proudly flying an ensign from the wings of a British *Large America* boat.

Sheds and Slipways at Felixstowe.

Above: Felixstowe flying boat station where the Felixstowe types were developed under the direction of John Porte.

Above: A Large America flying boat on a slipway at Felixstowe. (AHT AL0772-014)

Below: 1918 saw the Flying boats adopt colourful markings as displayed by the two Large Americas illustrated here.

1. The Development of the *America* Flying Boats

Above: Lt John Towers, USN, Glen Curtiss and John Porte (late RN) pose with the original *America* at Hammondsport some time after its official launching. The large enclosed cockpit with sliding windows is clearly visible in this photograph.

It is often stated that no US designed and built aircraft saw active service in World War I. What is not generally recalled is that US Curtiss flying boats were in use by the Royal Naval Air Service (RNAS) of Britain before the US entered the war. These flying boats carried out anti-submarine and anti-Zeppelin patrols over the North Sea and were frequently engaged in combat with German floatplanes.

On Bleriot's successful flight across the English Channel, the British newspaper publisher, Alfred Harmsworth, (better known as Lord Northcliffe), is reputed to have stated, "England is no longer an island". He was aware of the importance of aviation and the effect that German airships could have on England in the event of a war. He foresaw that the flying boat offered a solution to the problems with operating long-range aircraft in those days of low power engines. On April 1, 1913, Northcliffe's London Daily Mail announced a prize for the first flight across the Atlantic Ocean. The rules allowed 72 hours for the flight and the aircraft would be allowed to land for repairs and refuelling. The flight crew could board a surface vessel as long as they recommenced the flight no further advanced than when they landed. The flight could be from anywhere in the British Isles to or from anywhere in North America. It was to be hoped that the prize would promote real progress in aircraft design.

Glen Curtiss was the foremost builder of seaplanes at the time and he was approached by Rodman Wanamaker,[1] heir to the Philadelphia mercantile fortune, to build an aircraft to make the flight. In September 1913, Curtiss left for Europe to make arrangements for the licensed sale of his aircraft. In England he met John Cyril Porte, late of the Royal Navy. Porte had been invalided out of the submarine service with tuberculosis in 1911. Porte had been invited by Captain Ernest C Bass[2], a wealthy young Englishman, to join a syndicate to build/market Curtiss flying boats in the United Kingdom (UK). Bass gained the sole agency for Curtiss flying boats and engines in Europe, and in February 1914, transferred this to the White and Thompson Co. The latter set up a flying boat school and Porte was appointed flying instructor. It appears that Curtiss was impressed by Porte's flying of the Curtiss boat.

John Cyril Porte was born on February 26, 1884, at Bandon, County Cork Ireland. The son of an Irish clergyman, the vicar of Denmark Hill, he was the fourth child of seven children. Educated at private schools he joined HMS Britannia, the Royal Naval College at Dartmouth, at the age of 14, leaving

Above: John Porte in RNAS uniform. Porte was to have a large input into the design of the *America* to make it a practicable, usable flying boat.

at the end of December 1899. He then went to sea on HMS St George in January 1900. He remained at sea in various ships until February 1902 when, as a Sub-Lieutenant, he went to the Royal Naval College, Greenwich. In January 1904 he left Greenwich for the college at Portsmouth, leaving here in March the same year. Returning to sea in *HMS Royal Oak*, he remained at sea until Christmas 1904.

In February 1905, the same month he joined the Submarine Service, he was promoted to Lieutenant. He remained in the Service until August 1910. He served on the submarines *B3* and *C38*. It was while in the Submarine Service that he became interested in aeronautics. Returning to sea again in August 1910 on the Duncan, he became sick in October 1910 having contracted pulmonary tuberculosis. Given a year's leave, he was then placed on the retired list in October 1911.

Above: Glenn Curtiss, in his trademark cap, was the foremost builder of flying boats at the time but, like many early designers, worked from practical results and not theory.

Porte was a fine naval officer and his association with the submarine service is common amongst many of the early naval aviators. After three or four months in Switzerland he returned to the United Kingdom and began a career in aviation. He had built several gliders while stationed at the Submarine Depot, Portsmouth, and also purchased a machine, a little monoplane of the Santos Dumont Demoiselle type.

Porte learned to fly at Reims, France, and gained French Aero Club Certificate No.548 on July 28, 1911, flying a Deperdussin monoplane. He had a successful racing season and he started an aviation syndicate with an Italian, Mr D. Lawrence Santoni. This syndicate was later transferred into a company, the British Deperdussin. Aeroplane Company Ltd. The company ran a flying school at Brooklands and later at Hendon and were agents for the French Deperdussin company. Porte states that during this time he was working with aeroplanes and seaplanes.

In the latter part of 1913 he joined the firm of White and Thompson of Bognor. He worked with them for only a few months as Assistant Manager. While with this firm he first saw the Curtiss flying boat at Brighton and met Glenn Curtiss through his friend Captain Bass who had bought one of the Curtiss boats. Bass convinced Porte to join a syndicate to build/market Curtiss flying boats in the United Kingdom. Bass gained the sole agency for Curtiss flying boats and engines in Europe, and in February 1914, transferred this to the White and Thompson Co. The latter set up a flying boat school and Porte was appointed flying instructor. It appears that Curtiss was impressed by Porte's flying of the Curtiss boat.

Above: Porte poses with one of the hulls of the America flying boat at Hammondsport. The lower wing roots were built into the hull. The engine nacelles were braced to the top wing centre section to form a rigid structure.

Below: Moving the *America* hull out into the open for final assembly.

Right: Assembling the *America* in the open. Note the wall sided hull.

Right: Preparing for the launch of the *America*. The wing extensions have still to be fitted to the upper wing. The red painted machine made an imposing sight for the crowd of onlookers.

Porte was "very struck" with the possibilities of the Curtiss F-Boat and flew it several times, including times when Curtiss was present. It was the first time Porte had seen a flying boat. It "was a boat with wings on it rather than an aeroplane with floats on it and was therefore it had much greater seaworthiness and much greater stability."[3]

Curtiss had brought a 1913 Model F flying boat with him to demonstrate in Europe. This boat differed from the "standard" F-Boat in that the ailerons were attached to the trailing edge of the upper wing rather than mid-wing as on most other F-Boats. The English F-Boat was detailed in Flight magazine of 11 April 1914. The flying boat was criticised as it was considered structurally weak and poorly constructed.[4]

Norman Thompson also considered the Curtiss to be poorly constructed and set about overcoming these deficiencies in his design of the White and Thompson No.2 Flying Boat. Bass crashed his Curtiss boat and it was reconstructed with wings of RAF 6 section. Fitted with a 100 hp Anzani radial engine its performance was markedly superior as compared to that with the original Curtiss wings. In this guise it was known as the

Left: The *America* fully erected and ready for the official launching ceremony.

Below: No one had a British flag to match the US flag so Porte stuck a British postage stamp onto the hull. The name *America* is not yet painted on the hull.

Bass-Curtiss Airboat.

Porte contends that he was one of the first persons to talk seriously about flying the Atlantic as far back as 1911. He thought that the flying boat was the answer to the task of flying the Atlantic. He eventually saw Mr Tuchy of the New York World in London and discussed the idea with him. This scheme nearly came about but fell through at the last moment. He saw one of the directors of the English Daily Mail and shortly thereafter they made their famous offer of a prize for the first flight across.

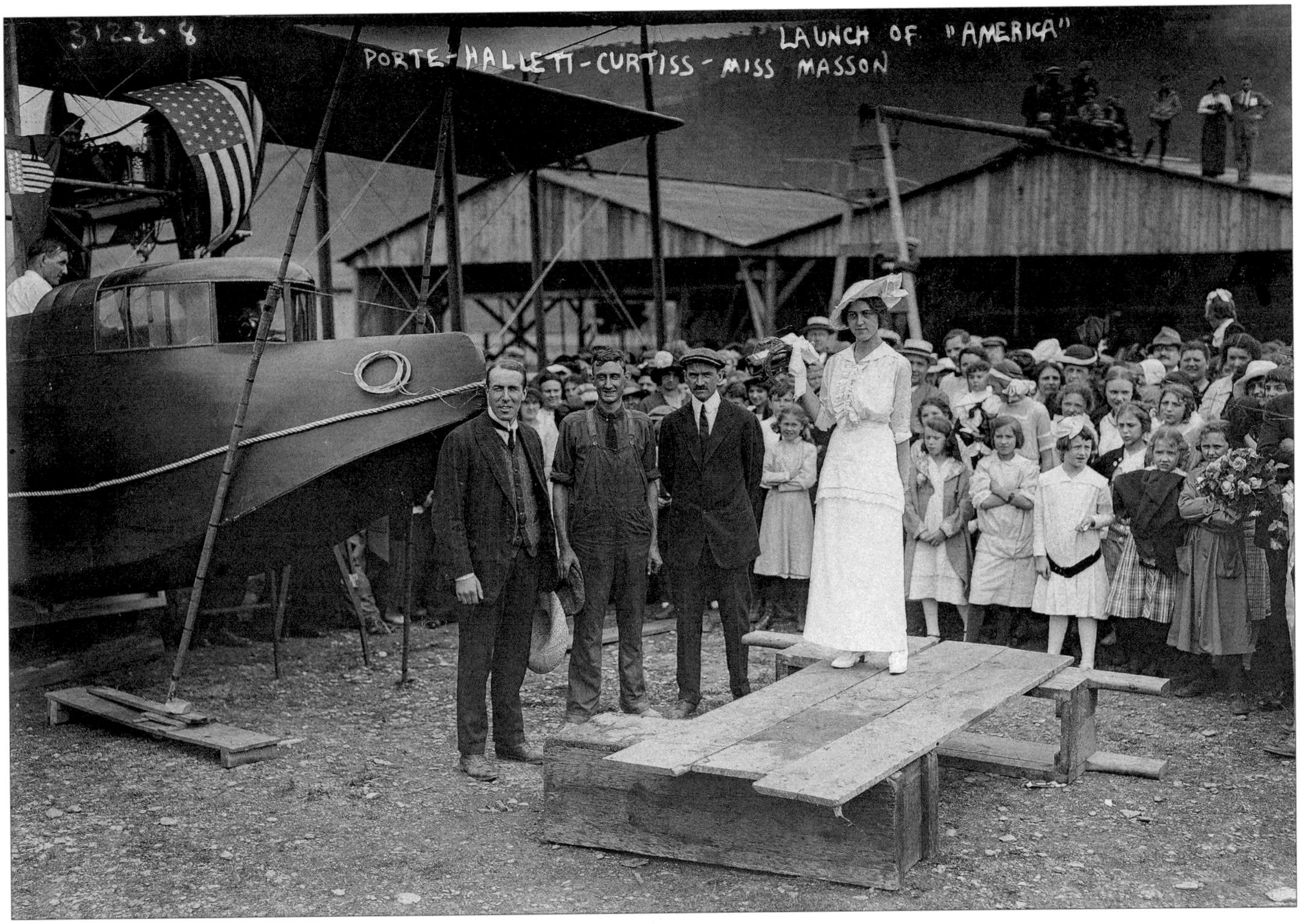

Right: The official launch of the *America*, 22 June 1914. Miss Masson stands with the appropriate bottle of Hammondsport champagne on a makeshift platform. The American flag and Aero Club of America flags on the boat are those presented by the Club to Curtiss when he won the first Gordon Bennett Aviation race.

Porte then obtained an introduction to Rodman Wanamaker in February 1914. Wanamaker agreed to finance the scheme using a Curtiss machine. Wanamaker's agents had already been in touch with Curtiss. Porte only stayed ten days and then returned to England.

Prior to his trip to America Porte had gone to Paris in December 1913 where he again met with Curtiss who was showing a boat there. Porte recalled that "Curtiss took a great liking to me and was deeply interested in my ideas for the development of his flying boat and we got on extremely well together." Curtiss and Porte discussed the Atlantic flight and Porte agreed to go to America "as long as my travelling expenses and one hundred dollars a week were paid for my living expenses." This amount was eventually paid by Wanamaker.

It was originally suggested that the machine for the Atlantic flight be a land machine with a 200 hp motor. When he realized that such a motor was not going to materialize, Porte states that he went back to Curtiss with the suggestion that "a large boat with 2 of his engines of 90 horse power each" would be the best machine for the proposed flight.[5] Glenn Curtiss testified that Porte had suggested a floatplane with two cylindrical floats that would also serve as fuel tanks. Whatever the truth of the matter, the machine finally agreed on was a twin-engined flying boat.

In April 1914 Porte returned to America with the intention of "helping to design the machine and prepare the boat and make all necessary arrangements for flying" the Atlantic.[6] "In fact in this particular machine the improvements were entirely my idea…this machine was built under my supervision and I also made a number of suggestions with regard to the alterations to the hull which were carried out and my suggestions were taken up by Mr. Curtiss."[7]

Porte had worked closely with Norman Thompson on the work of improving the Curtiss F-Boats. The experience was his introduction to his practical knowledge of flying boat construction, and he took this expertise with him to America where it enabled him to be of practical assistance in the work of designing the *America's* hull.[8] Casey[9] gives credit for the design of the America boat to B. Douglas Thomas, an Englishman that Curtiss met on his European trip in 1913. Thomas designed the Curtiss Jenny and it would be logical that he would have had a great input into the design of the boat, especially the

Above: The *America* taxiing.

Left: The *America* was a "wet" boat on the water as it is shown taxiing well below the hump speed.

Below: The *America* could get off the water with its original wall sided hull. Note the two different design radiators.

wing structure.

Wanamaker wanted a mixed American and British crew as he wanted the flight to celebrate 100 years of peace between the USA and Britain. Curtiss was so impressed with Porte that he had invited him to join the crew of the proposed trans-Atlantic boat as pilot.[10] Lt John H Jack Towers, USN, was suggested by Curtiss as the American half of the crew. Towers was a pilot as well as an expert navigator. Towers had flown the Navy's first hydroaeroplane, the Curtiss A-1, and had carried out experimental work with Curtiss on Naval aircraft. The USN had difficulty in accepting that a serving officer would be co-pilot to a British citizen, and the American crewmember was not resolved when Towers was ordered back to duty, sailing for Mexico in April 1914.[11]

The Wanamaker flying boat was erected in on the shore of the lake as no building in the Curtiss factory was large enough to house it. The biplane flying boat was a handsome craft with a single step to the hull and of narrow beam. The hull was

Above: The name must have been applied to the hull at the launch or immediately thereafter as the boat still carried the flags from the launch ceremony. Curtiss, in cap, talks to some locals on the left.

built on a framework of white cedar and ash following boat building practice with stringers and formers to give the hull its shape. The skin comprised longitudinal ½ inch strips of spruce secured with brass screws. The V-bottom of the hull had two layers of ¼ inch planks laid at an angle to each other, the layers separated by a layer of cotton fabric set in marine glue. The whole was covered with red Japanese silk and varnished. The wings were in seven sections; a centre section; four main panels and two overhangs at the extremities of the upper wing. They were covered with red silk. In the event of a mid-ocean landing the wings could be jettisoned. The rudder was attached to the rear of the hull which terminated in a vertical knife edge. The lower section of the rudder was metal bound oak for steering in the water. The upper section was constructed of ash and covered with fabric. Power was provided by two 90 hp Curtiss OX V-8 water-cooled engines mounted between the wings turning eight-foot counter-rotating pusher propellers. There were two bays of interplane bracing outboard of the engines. The upper wing had long extensions that were braced from king posts on the upper wing set above the interplane struts.

Four watertight bulkheads plus four watertight tanks in the stern were to provide buoyancy in the event of a forced landing. Three tanks were mounted forward of the step and allowed for 500 gallons of fuel. Fuel was pumped from the main tanks to a gravity tank in the upper wing by a rotary gear pump. An auxiliary hand pump was fitted for emergency use. Unusual for the time "over the body of the boat is a permanent cabin top fitted with celluloid windows on top, in front and on both sides, forming an enclosed cabin and pilot house." The pilots sat side by side on a bench seat with dual controls. The wheel controlled the rudder and each aileron was independently controlled by the foot pedals. The "horizontal rudders" (elevators) were controlled by lateral movement of the control column. This arrangement was incorporated due to the litigation being undertaken by the Wrights over patent infringement. A mattress was situated in front of the fuel tanks so that one crewmember could rest at a time. Directly in front of the pilots were the instruments including tachometers, aneroids, wind speed gauge, inclinometer, fuel and oil gauges, compass, etc.

The triangular fin was of low aspect ratio with a rounded rudder of pleasing shape. The tailplane was mounted high up on the fin.

On June 19, the name for the new flying boat was announced. It was America. In 1910 Walter Wellman, together with five crew, had tried to fly a hydrogen balloon across the Atlantic. Their effort came to nought when they were wrecked in a gale and miraculously; all were rescued by a British mail packet. The balloon had been named America.

The 16 year old daughter of Curtiss' friend, Leo Masson,

Left: Porte glances at the photographer as the *America* is lined up for the waters of Lake Keuka. Note how the forward upper hull decking was a fabric covered wooden framework.

Below: The *America* enters the water, probably for the first time.

had been selected to officially christen the boat. On the 22nd the ceremony took place. Unfortunately a British flag could not be located. After two failed attempts a sledgehammer was used to break the bottle of Great Western champagne and the boat was pushed into the water. It was too late to fly and it was to be the next day when the engines were run up and Curtiss and Porte took the aircraft up at 3.00PM for its first flight.

Who Designed the America?

In June 1923 the Royal Commission on Awards to Inventors considered the Curtiss Corporation's; Mrs Porte's and the Norman Thompson claims at the same time. As the Curtiss attorney had intimated that Porte had no input into the design of the America flying boat, Glenn Curtiss was called as a witness. Mrs Porte's representative questioned Curtiss in depth as to the events that led up to the successful conclusion to the America boat as purchased by the British. Curtiss insisted that he was the designer but after evidence was produced in the form of correspondence between Porte and Capt Creagh-Osborne, RN, Curtiss agreed that Porte did have some input into the design of the America. He stated that "There is no question but that Colonel Porte's experience was of value and these matters were determined by experiment there and then."

Sometime in April 1914 Porte wrote to Creagh-Osborne, who was to be sent to the USA to keep the Admiralty up to date on the progress being made at Hammondsport. According to Glenn Curtiss' testimony, he was only aware that Porte had requested Creagh-Osborne as he was an expert on compasses and Porte wanted "to get him over there to help arrange and test the compasses." In his letter to Creagh-Osborne, Porte wrote that "he had not been able to arrange anything definitely yet as the date of the flight seems uncertain, but during the past

Above & Below: A crude arrangement to extend the planning bottom width is shown here.

week we have altered the whole scheme, chiefly through my persuasion and arguments so I think by the end of this week we may know something pretty definite all round...We have now designed a real sensible machine which I hope will be able to rise decently from rough water and carry about 22 hours fuel supply."

In a further letter to Creagh-Osborne, Porte left a detailed account of the experiments undertaken to bring the America to a standard that would be capable of undertaking the transatlantic flight as follows:

Result of Trials to Date, July 9th

"Directly the America was taken out for the first time it was seen that she had insufficient planning surface and would only get off the water with about 3,500 pounds total weight. Small submerged fins of about one square foot each were fitted each side of the step, and proved f no value at all. A radical change was then made and large fins of triangular (sic) cross section, were fitted each side of the hull in front, the bottom sloping slightly outward and upward. These extended 2' on either side with the back edge at the step. It was at once seen that there were a considerable improvement. The total load raised was 4,400 lbs, but this was the limit.

The main step was then extended back some 40" and tried again. This proved useless as the machine persisted in violent porpoising. Extensions were then fitted to the back and sides of the fins, but made of flat boarding. These did not seem to have any effect, probably due to their not assisting at all in the floatability of the machine. Two flat pontoons were then fitted underneath the first strut outside the engine section.

These pontoons were ten feet by three by eight inches/deep, the bottom being three inches above the bottom of the boat. These again proved of considerable assistance, the machine raising a total weight of 5,000 pounds off the water.

However, all these fittings were of too complex a nature to be serviceable so the whole lot was discarded and a false bottom 14 feet wide was built underneath the boat, this false bottom being made of flat planking slightly sloping upwards on each side of the boat. This again proved entirely useless, due no doubt to no increase in the floatability of the boat. This false bottom was again discarded and submerged fins substituted, these extending from just in front of the step at right angles to the boat slightly sloping upwards; these were six feet by a foot and a half. On trying these the machine immediately came to the surface bringing the fins out of the water with the result that he machine immediately dropped back again to her original position without having attained a speed greater than twelve or fifteen miles an hour.

A second fin was mounted underneath this, shorter and a foot lower, being beneath the bottom of the boat, the only result being the machine rearing right up on the water on the lower fin and repeating the operation of porpoising and falling back again. These fins, although they might eventually be made to work on a smooth lake or river, it is doubtful if they would ever be of any value in a seaway. After this the fins were cut off and pontoons reverted to, these are still on trial."[12]

The hull was originally wall sided and the seaplane had difficulty in getting off with this hull. Experiments were carried out to overcome this problem. Curtiss documents give the following diary of events:

June 23 20 minute taxi flight at 3.00PM. 4.06PM Porte made second flight.

June 24 Hydroplane boards added to hull did not work as aircraft could not lift required load. Men worked all night to change boards. Porte made night flight at 8.00PM, carrying 500 lbs more than that of the 23rd.

June 25 Hull changes worked well. The boat raced a Curtiss M-boat and Curtiss in his hydroplane boat Skimmer. The hydroplane boards were changed to a larger size (24 inch) by the men working all night in order to allow the aircraft to carry a 1,400 lb load.

June 26 No flight.

June 27 Seven men carried aloft.

June 28 Storm threatened aircraft.

June 29 One flight with eight men and a second with ten men (reporters). Removed tail hydroplane.

June 30 Three flights made. Confirms can carry over 4,000 lbs but needs safety factor.

July 1 Two 1½ minute flights. Side fins lengthened and special Olmsted propellers were fitted. In this form the aircraft met its specifications. Curtiss was not satisfied as Olmsted propellers were prone to splitting.

July 2 Olmsted propeller tests. Langley pontoons added for tests. New bottom to be placed on boat – two feet deep loaded.

July 3 Test of new hydroboards (so called "barn door bottom"). No flight.

July 5 Flight tests on one engine. First one then other. 3.00PM flight with seven passengers (4,500 lb load).

July 7 Porte flies America.

July 8 Langley pontoons placed back on boat.

July 15 Rebuilding hull after water trails.

July 16 Sea sled hull repaired and tested on water.

July 18 Test failed as nose rose and tail went down into water. Possibility of third engine considered. Also the engines did not develop their designed power, but only about 85 hp. Curtiss installed a third motor driving a tractor propeller on the wing centre section.

July 20 Third motor added. Changes in hull. Auto searchlight mounted with lamp in cockpit and small hand lamp for use by mechanic.

July 23 Flight with third motor. Must carry 2,289 lbs for flight across Atlantic, flew with total load of 2,654 lbs. Fuel consumption tests. Tip flew off Olmsted prop damaging wing rib and fabric. Repairs delay teat until evening. Bottom to be rebuilt before shipping to Newfoundland.

July 25 Trans-Atlantic flight delayed till October. Time required to rebuild bottom of hull and more test flights need to be made. On test it was discovered that the third engine's fuel requirements offset any advantage. 28 changes were made to hull during experiments and came back to second plan as the best method.

Port was to recall that the whole bottom of the hull was then rebuilt with fins extending from the stern (Porte must mean the bow. Author's note.) as far as the step on each side. The overall width being 8 feet and the hull proper only 4 feet. In cross section the bottom showed a slight (straight sided) V form from keel to fin edges and extending as far aft as the step. The tail being flat bottomed. This form of hull was embodied in all the *Small Americas*.[13]

It is to be noted that Curtiss carried out trials and modifications in the field in a crash program with a full sized machine rather than theoretical knowledge from model hull experiments. There was little knowledge available at the time relating to the problems that they were trying to solve. Porte would use the same methods when he returned to the UK and was placed in charge of developing an effective flying boat. As noted above, the trials on Lake Keuka confirmed that the America, the largest flying boat built up to this time, could not carry the required load for the trans-Atlantic flight. With full load it would not lift from the water. The bow would submerge when power was applied and a program began to add buoyancy and increase the planning area of the hull. Hydrofoils were tried at least two different locations before being discarded. Various other methods were tried including fitting a broad sea-sled type of device to broaden the hull. Later this was matched by fins/sponsons built into the hull. Also the engines did not develop their designed power, but only about 85 hp.

According to F.N. Hillier, who was with Porte "during those

Right: Close examination of the photograph shows that the *America* has one of the hull extensions fitted from the cockpit to the rear of the wings in a much neater, more workmanship arrangement. A Curtiss E-Boat sits on the shore of Lake Kekua behind the larger machine.

Above: This photograph shows one of the methods tried to give the hull better hydrodynamic qualities. The opening in the top of the hull behind the cockpit to give access to the engines is well shown.

days of fruitless preparations at Hammondsport, New York, during the summer of 1914," up to the end of July, "when we all packed up to come home in view of the imminent outbreak of war, it had been impossible to coax the aircraft off Lake Keuka with any better load than two bodies and a few gallons of petrol."[14]

On the declaration of war, Porte immediately returned to England. "In fact I actually sailed from America on the 4th August one and half hours after war was declared. I was so rushed that I left a lot of my clothes and papers at Hammondsport. I had a long interview with Mr. Curtiss in New York just before I left to come back."[15] He asked if the two boats would be available if the British Admiralty wanted them and Curtiss agreed that they would be available.

As soon as he arrived he volunteered his services to the Navy and, although still suffering from tuberculosis, he was accepted and appointed a Squadron Commander, and took up duties at Hendon. He saw Commander Murray Sueter[16], the head of the Air Department of the Admiralty and Porte informed him that the two America flying boats were available for sale in the USA. Porte did not know the price of the boats but knew that they would "cost a great deal of money."[17] Sueter recognized the value of such machines for the new type of warfare that was likely to be waged against German submarines around the British coast, and asked Porte to find out the cost of obtaining the two boats and the cabled prices were £6,000 and £4,500.

In a letter to Sueter on 14 August Porte wrote that the "hull is built very strongly and should be capable of standing a good sea. She will arise from the water with a great facility except when heavily loaded when it is necessary to have a smooth sea. The machine is very steady in the air and specially built to facilitate navigation, the compass behaving well. This machine should be finished this week, and I have no doubt it could be shipped in about three weeks."

Above: Installing the third engine on the upper centre section. The hull is showing signs of wear and tear.

Below: The upper wing has been cut back to the front spar in order to provide clearance for the third motor's propeller.

Right: The crude planning bottom extensions tried on the *America's* hull are clearly seen.

Below & Bottom: With the third engine the machine could not carry enough fuel for the proposed trans-Atlantic flight.

Above & Below: This was the final hull with fins and was how they were received by the RNAS.

It is assumed that Porte was referring to the changes that had to be made to the original America rather than a completely new boat.[18] He continued…

"A duplicate is about three parts built and could be finished in about three weeks. If it is desired, these machines could be purchased privately and shipped via Halifax, but there seems some doubt whether this course will be necessary."[19]

Sueter considered that "these machines would be used for operations in the North Sea as they can lift a large weight, either in the form of fuel for scouting purposes, or in the form of explosives for offensive purposes." He considered that the price asked was fair and that while no money was available for the purchase, it was a small amount, and he recommended that the two machines be purchased for £9,500. Delivery would be

Above: The final form of the rebuilt *America* hull showing the addition of what were termed "fins".

in New York, packed ready for shipment as Curtiss "refuse to consider the question of shipping the machines."[20]

The initial price asked was considered too high and Porte was asked to negotiate a reduction. Curtiss dropped the price to $25,000 for the first machine and $22,500 for the second one, which Porte considered a "considerable reduction." This was accepted provided that Curtiss deliver both complete and properly packed for shipment. It was stated that the names "AMERICA" or "TRANSATLANTIC" were not to be used. The reason for this is not stated, but may have been an attempt to disguise the shipment from German agents. If so then they were not successful. The Admiralty's General Agent in New York wrote that while the greatest secrecy should be observed, the shipment had been well advertised in the UK and the British Consul General "has been mixing in this quite a great deal." The small town of Hammondsport where the machine was shipped from meant that it was impossible to keep its movements a secret; also a special train had to be employed. US newspapers commented on the acquisition and how it may have international implications due to the war.[21]

The Admiralty purchased the two boats on Contract No.C.P.52840/14. The first was shipped via the *Mauretania* as deck cargo under a dummy bill of lading. The Germans protested but the US Secretary of State declared that aeroplanes were not contraband of war. Designated H-1 in British service they received serials 950 and 951.[22] The first arrived at Felixstowe in October 1914, the second about six weeks later via the infamous SS *Lustinia.*[23]

They were taken directly to RNAS Station at Felixstowe and erected. Porte went to Felixstowe to see the boats when they arrived but was not involved with them at this time as he was in charge of Hendon aerodrome. "I was to test the first one and Commander Suter (sic) himself came down to take a flight in it and examine it himself. I took him for a short flight, that was at the beginning of November 1914 and he returned to London." On arriving in London Sueter reported to the First Sea Lord, Winston Churchill, praising the boats saying that he considered them excellent. Apparently Churchill was impressed as he wrote on the report: "Order 12 of these at once."[24] According to testimony given to the Royal Commission of Awards to Inventors post-war, none of the *Small America* boats ever made a patrol, the 56 supplied by Curtiss being used only for training in sheltered waters. Other sources state that patrols were made but being experimental they may not have been officially considered as an active service patrol?

Originally these boats had lifting tails which made them easy to get off the water with a full load, but they were not very stable in the air. To correct the instability the weights were rearranged and the tailplane incidence decreased. This made them stable fore and aft. In rearranging the weights the engines were changed from pusher to tractor configuration. This improved their flying qualities but they were nothing

Right: Although supposedly based on the Curtiss design, the Italian Bossi *America* type flying boat bore little resemblance to the original design.

like as good in getting off the water. Anzani engines were substituted for the Curtiss engines as the latter had proved very unreliable. Although the boats were still satisfactory in getting off in smooth water, in a sea way they were very wet. Water was frequently thrown over the engines in sufficient quantities to cause them to misfire and prevent getting off.[25]

Porte recommended that the boats be built in the United Kingdom rather than being purchased abroad. Eventually an order for eight was given to the Aircraft Manufacturing Company (AMC) at Hendon (1228–1235), and four were ordered from Curtiss (1236–1239). As explained hereunder, there were some dubious dealings undertaken with respect to this order with the AMC.

Four Curtiss Transatlantic "America" type Seaplanes were ordered at a cost of $22,500 each together with 20 Curtiss OX motors (ten LH and ten RH) at $2,800 each for a total contract price of $146,000.[26] The boats were to have "the Deperdussin type of control with ball bearing pulleys" fitted in each machine. Lyman J. Seely, the Curtiss Company representative in the UK, sent a Telegram to the Curtiss Company at Hammondsport noting that the official acceptance of the tender for "four H. Oats twenty motors" would be in the night's mail. In this telegram Seeley states that the specifications "demand boats with long tail hulls like *America second*_(Author's italics) with wings attached at center (*sic*) by sockets and pins for quick demounting hulls tighter bottoms braced and strengthened between wings beams and bottom."[27]

The Admiralty also asked that the question of providing folding wings be considered but this came to nothing. The four Curtiss built boats arrived in the spring of 1915 and were sent to Felixstowe. They were received long before those manufactured in England were completed.

These latter boats should have been produced early in 1915 but were delayed for a variety of reasons. They were constructed under Porte's supervision at Hendon, the local manufacturers were allowed to measure and take details of the construction of the Curtiss boats.[28] The hulls were built by Saunders at Cowes using their patented Consuta form of construction. The straight lines of the Curtiss boat were replaced by curves, the planking being sewn with stranded copper wire. They were not a success mainly due to their form of hull construction, the copper stitching breaking through the planking. The work done with these boats tie the development of the Curtiss machines into the following Felixstowe F-boats and their story is detailed in Chapter 2.

Marketing of the *America* Boats

That it had not started for its momentous voyage from St. John's, Newfoundland, to the Azores Islands under the guidance of Lieut. John Cyril Porte, R.N., at the time war was declared was due to so small a chance as the accidental stripping of the copper sheathing from the special propellers he was to have used.

So ran Curtiss advertising following the outbreak of World War I as Curtiss tried to market the *America* flying boat. From the above there can be seen a distortion of the truth in that the boat would not have embarked on the trans-Atlantic flight until after more testing and redesign.

In his advertisements Curtiss noted that the machine could be used "either as a sporting vehicle or for naval use…Her weight-carrying capacity and her steadiness make the mounting of a gun of considerable size, the use of light armor (sic) plate, and her equipment with long range wireless apparatus entirely feasible." As a civilian machine it could carry up to a dozen passengers or for long range cruising it would be fitted with folding berths for two or three men.

As far as is known the only *America* boat ordered by other than the British was one ordered by the Danish Ministry for the Navy in the autumn of 1916. Lt (jg) W. Capehart of the

Flying Reproduction of the America Flying Boat

On this page are color photographs of a flying reproduction of the original *America* flying boat. (Courtesy D. Ostrowsky)

This Page & Facing Page: First flight of reproduction America, Hammondsport. (Courtesy D. Ostrowsky)

Curtiss

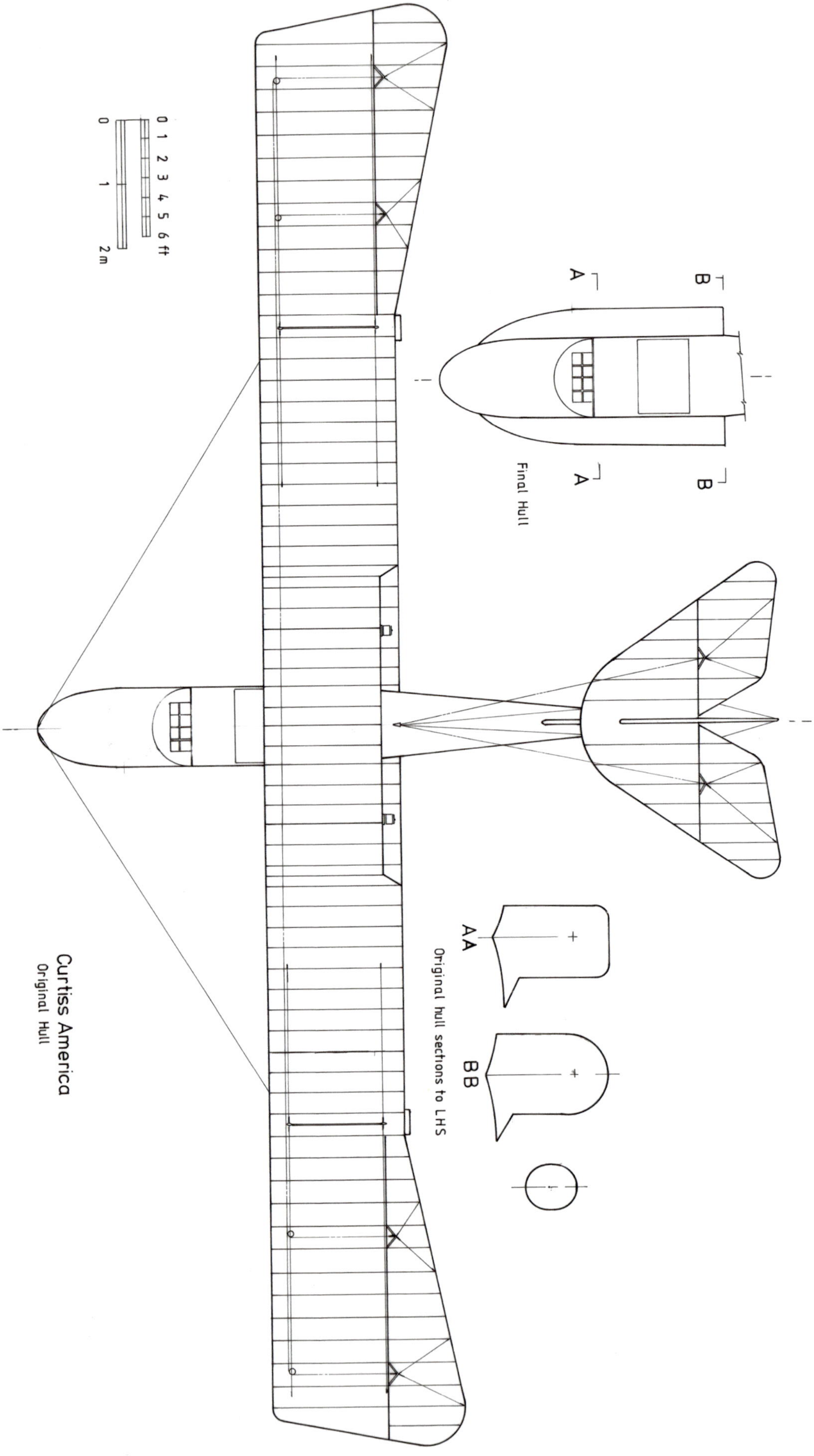
0 1 2 3 4 5 6 ft
0 1 2 m
B
A
Final Hull
B
A
AA
BB
Original hull sections to LHS
Curtiss America
Original Hull

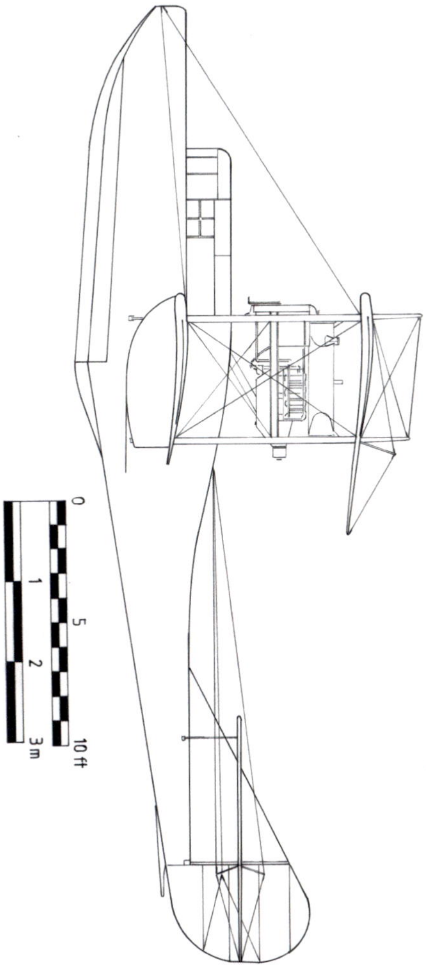
0
1
5
2
3 m
10 ft

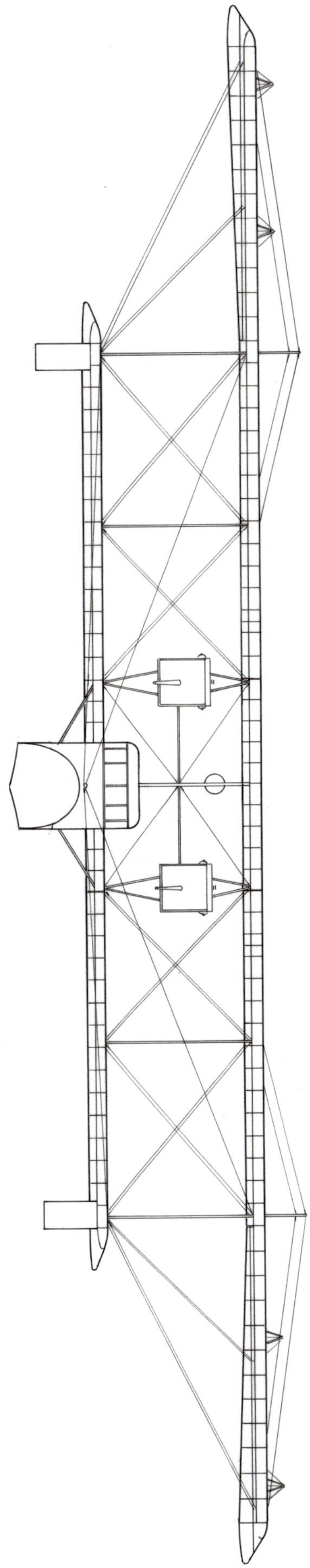

Above, Below, & Bottom: The Danish H-4. This machine did away with the enclosed cockpit and had a large rounded balanced rudder while the rear hull was flat bottomed.

Above: RNAS Curtiss H-4 3549 has 100 hp Anzani radial engines installed. The reason for the interesting colour scheme applied to this boat is not known. 3549 was delivered to Felixstowe for erection on the last day of July 1915. Marked for deletion in December 1917, it was presumably rebuilt as it was at the Felixstowe Seaplane School by February 1918. It survived until deleted the following August.

USN Bureau of Construction and Repair, reported on the boat built for the Danish Government in November 1916.

Capehart wrote that the H-4 type P.B. (Patrol Boat) flying boat had a windmill pump that forced fuel from the 200-gallon tank in the hull to the 12-gallon gravity tank that fed both motors. The auxiliary oil tank contained 8 gallons for each motor. Zentih Duplex carburettors were installed. Each motor had an independent hand throttle and switch.

He did not like the "big overhang of the upper wing for structural reasons". He considered that the lack of leverage for the tail controls was "evidently made up for by size." The vertical fin and horizontal stabilizer looked like "the biggest I have ever seen." He also considered that the boat had an usual shape and referred to the photographs in his report.[29] It is evident that he was not impressed with the machine.

At 2PM on 10 November, Carlstrom, the Curtiss test pilot, made a short solo flight. The machine appeared to get off satisfactorily but it was immediately evident that the machine was tail heavy or else the setting of the tailplane was out. In order to get back onto the water without tail sliding or stalling, Carlstrom had to jolt the nose down by sudden and frequent bursts of power. He was finally able to make a 12 minute flight that afternoon after several attempts to get the boat into flying shape. He had 700 lbs of weight in the bow to overcome the tail heaviness.

The next day it was discovered that the hull had a great deal of water at the tail. This was the cause of the extreme tail heaviness the day before. With the boat dry and 335 gallons of fuel in the main tank, the boat was balanced with Carlstrom alone and 150 lbs ballast in the bow. When he came back from his first flight of the 11th, Capehart and Lt Scofield joined him

Above: The large rounded, balanced rudder and flat-bottomed rear hull of the Danish H-4.

for a timed flight. Scofield sat back in the body of the hull near the main fuel tank. The average speed over the one mile course was 66.6 mph. On trying to climb, 300 feet was reached in the first minute, then the port engine gave trouble and the boat came down.

At slow speed in rough water, the boat would plow badly. It planes readily, gets off readily, and lands readily. I could see no particular difference between its action and that of the smaller flying boats which I have flown at Pensacola. I do not consider it would be a satisfactory machine for our use.

Capehart noted how hard Carlstrom had to work the controls. He considered it would tire a man in about 15 minutes. "It was impossible for him to turn the rudder far enough for it to be of any great use." Carlstrom would run over the course, land, turn around on the water and then take off again and run back over the course in the opposite direction. Capehart considered it too dangerous to attempt a slow speed run under these conditions. "It seems likely that "Servo" Motors would be necessary for the aileron controls of a big boat as well as the rudder controls, or balanced controls."

The following notes were made by Capehart:

- Do not like such big overhang of upper wing for structural reasons;

Above & Below: 1232 was an Aircraft Manufacturing Co (AMC) version of the H-4 *Small America* flying boat. Delivered to the Killingholme Seaplane School in on 1 August 1916, 1232 was not deleted until April 1918. The hull appears to be varnished with clear doped fabric surfaces. The cockades are large and overlap the ailerons. Note the Lewis gun on the top of the cockpit roof.

- Tail control wires lead out of Boat body on way aft just inline with propellers;
- Clearance between the Propeller blades and control wires about 10 inches,
- Motors were started by hand cranks every time, but with difficulty;
- The Boat body is relatively short;
- Lack of leverage of tail controls evidently made up for by size;
- Vertical tail fin and horizontal stabilizer looked like biggest that I have ever seen;
- Boat has unusual shape;
- Do not like structural details of tail controls, horizontal stabilizer, etc.; should hate to have to pull out of steep dive suddenly;
- Horns for control leads to rudder are very short, giving poor leverage, – 6".[30]

1st Lt Asger E.V. Grandjean, later commander in chief of Danish Naval Aviation, was present when the aircraft was

Above: This H-4 with a Saunders built hull (probably 1232), displays the difficulty of manning the Lewis gun that was carried on a pivot mount in front of the cockpit. The Lewis is an infantry model with a bag attached to catch spent shells from hitting the propellers. Photograph believed taken at Killingholme.

Above: AMC H-4 1234. This photograph was probably taken at Killingholme where the machine was allocated in October 1916. The upper surfaces of the wings appear to be a dark colour while the underwing cockades unusually show a white outline. 1234 was written off after a crash on 23 July 1917.

Right: Curtiss H-4 3585 has its serial in white on the light hull. The Anzani radial engines do not have the fairings seen on 3555. Delivered to Felixstowe for erection as were all these boats, 3585 served from January 1917 until it was deleted in January 1918 as beyond repair.

Above: Curtiss H-4 *Small America* 3555, one of 50 boats ordered from Curtiss that were mostly re-engined with 100-hp Anzani radials. The serial was made from a stencil with the serial numerals being the same colour as the hull. Note how the cockade overlaps the aileron. Delivered to Felixstowe in September 1915, 3555 served at Calshot from December. It was written off on 9 February 1917.

Above: 3592 was another of the Curtiss built *Small Americas* that served so well as flying boat trainers. This machine was at the Felixstowe Seaplane School by 29 September 1917, and was written off, together with 3584, 3586, 3591, 3593, and 3594, in the week ending 29 August 1918.

test flown on Lake Keuka at the beginning of 1917. The original 100-hp Curtiss OXX-2 engines could not meet the performance specifications and were replaced with engines of higher power. However, these did not improve the performance of the aircraft sufficiently and the Danish authorities refused acceptance and the contract was annulled. The funds were used to purchase a number of 100 and 200-hp engines that were installed in Danish made *Maage* (Gull) flying boats.[31]

There is mention of a boat for Norway in USN documents. Photographs in the file show an America type boat with conventional tail surfaces, a balanced rudder and pusher engines.[32] A Memorandum from J.W. Scott of the Curtiss Aeroplane Co, to Hunsaker, noted that a H-4 being constructed for Denmark would be ready about 8 November 1916.[33] The Memorandum continues…

"A somewhat similar aeroplane built for Norway was inspected by representatives of the Bureau(s) of Construction and Repair and Steam Engineering at Hammondsport, N.Y., recently, and was found badly out of balance and recommended as unsuitable for purchase."

Above: An original *America* hull for comparison. Note the sharp bow projection.

Above: Standard Curtiss H-4 *Small America*. Anzani radials replaced the Curtiss engines.

Above: Compare this hull with the two above. The fins start at the nose and the cockpit has transparent panels in the roof.

Left: H-4 taxiing.

Above: Although the biggest flying boat in the world at the time of its launch, the *America* was actually quite small when compared to the H-12 that followed it into service, hence the name of the boats changing to *Small America*. The crew of the boats would be piggybacked by the wader dressed ground personnel out to or from the boat.

It is possible that the Danish machine was confused and referred to as one for Norway. The tests attended by Capehart may have been for the re-engined Danish boat. As with so many early aircraft, it is hard to be precise as to which aircraft is being referred to in surviving documents.[34]

In March 1916 Mark L. Bristol, the officer in charge of US Naval Aviation, noted that the large type of flying boat produced by Curtiss "commonly known as the American type, has received no real demonstration in this country in rough water." Curtiss was going to have one of the machines shipped to Hampton Roads where "they have a flying station" and then it would be possible to demonstrate the type in the open sea. "A number of these machines have been sent abroad, but exact knowledge as to their behaviour in the open sea is not available."[35] No documents relating to such a test have been found. The USN continued to develop the pontoon aircraft for sea use.

Curtiss H-4 *Small America* in British Service

The Admiralty used its seaplanes for anti-submarine patrols and mine hunting in support of the ship patrol as well as its responsibility for home defence. The Sopwith and Short floatplanes were next to useless for this work but were all that was available. It was felt that the Curtiss boats would give the Admiralty an effective weapon.

The first two boats supplied to the Admiralty were the original America boats. This is confirmed in US newspaper articles of the time. The first boat, 950, was delivered to Felixstowe on 13 October 1914, with the other, 951, following on 25 November.[36] John Towers, USN, was in the UK to report on British developments for the US Navy at the time and while "poking about Hendon…watched Cyril Porte test the two modified Curtiss America flying boats just brought over from Hammondsport."[37] Until the US entered the war on the Allies side, Towers received very little cooperation from the British. F.N. Hillier recorded that when he next saw Porte "at Felixstowe in the early summer of 1915, the America, in which Porte gave me a short flight, had two sister ships which were doing three-hour patrols over the North Sea."[38]

As noted above the next Curtiss boats purchased by the British (serials 1236–1239), were ordered from the Curtiss Aeroplane and Motor Co, Hammondsport, and were similar to the first two boats. Although they are referred to under the designation of H-4, Casey states that one of these was a sister ship to the America (an H-1) and the other three were H-2 boats.[39] The information supplied by Curtiss in its claim for compensation post-war indicates that these four boats carried the factory designation H-2. The British designation H-4 will be used here. The H-4 had a hull two feet longer than 950, the tail was more square and, unlike the AMC/Saunders hulls, they had a plain V bottom.

The first Curtiss machine (serial 1236) was delivered to

Above: This H-4 appears to have a varnished hull with white cabin roof. Wings appear to be clear doped.

Felixstowe on 17 January and the last in March 1915. The first AMC boat was not delivered until 20 June 1915.[40] In February 1916, C.L. Vaughan-Lee, the RNAS Director of Air Services wrote that *there had been some cases of Aeroplanes and Seaplanes on order whose delivery has been so long delayed that they are hopelessly out of date and already useless before delivery.*

Vaughan-Lee wanted such contracts cancelled on "the best possible terms". The AMC contract was one considered by Vaughan-Lee for cancellation as five of the eight Curtiss Boats ordered "have been overdue for more than one year and action is being taken to consider the cancelling of the outstanding portion of the contract."[41] Despite this the AMC contract was completed the last boat being delivered to Felixstowe in January 1917. Porte recorded that the "Boats were built under my supervision at Hendon. The hulls were built by Sanders (*sic*) at Cowes."[42]

On 12 March 1915, Felixstowe had a White & Thompson, two H-1 and four Curtiss built H-4 flying boats on strength as well as a number of Short floatplanes. The Curtiss engines were replaced by 150-hp Sunbeams or 125-hp Anzani radials as the Curtiss engines did not produce their required power and were considered unreliable. It was reported that the original boats could not get off the water with their Curtiss engines even when all excess weight was removed from the boats. It was found that the hull was weak at the step and would break at this point in a sea wave. Due to the very short fore-body and lack of buoyancy forward the boat was entirely unseaworthy. The hull was so weak that the tailplane would sag under its own weight and alter the trim of the boat once the machine got under way.

Despite its shortcomings the First Lord of the Admiralty pointed out that "the Curtiss machines, and in fact all aeroplanes and seaplanes delivered from America – must be considered as an addition to our forces. The machines must be taken as they are, and must be made the most of in spite of the defects."[43] As noted above, eight America type flying boats had been ordered from the Aircraft Manufacturing Co (Serials 1228 to 1235) under Contract C.P.65070/14. The first three British built H-4 boats were delivered from June 1915; however the latter five not delivered until 1916.

Curtiss received an additional order for 50 "H-4 Transatlantic Boats" in March 1915 under Contract C.P.01533/15 (Serials 3545–3594). As related below, these would be the aircraft from the order for 150 machines that Porte reduced to 50.[44] Records show that 31 boats on the order were delivered as complete machines, while the rest were delivered without engines for a total cost of $1,125,000. The same schedule shows that nine spare H-4 hulls, 18 sets of spare main planes, rudders and wing floats, plus nine tail units, were ordered under Contract No.C.P.105569/16/X, and were all

Curtiss America Specifications

Source	1. Flight	2. Curtiss Model H	3. Transatlantic Model Curtiss Flying Boat	4. America Flying Boat	5. Danish America Boat	6. British H-4	7. British H-4
Length o/all	34 ft	36 ft 9 in	34 ft	33 ft	37 ft 0½ in	37 ft 0½ in	37 ft 7¼ in
Length Hull	30 ft		30 ft				
Width hull	4 ft		5 ft (8 ft over fins)				
Depth hull	6 ft		6 ft				
Span Upper	74 ft	75 ft	76 ft	76 ft	75 ft 9¾ in	75 ft 9¼ in	75 ft 9¾ in
Span Lower	46 ft	48 ft	46 ft	46 ft	48 ft 0 in		48 ft 0 in
Chord	7 ft	7 ft	7 ft	7 ft	7 ft 21/32 in	7 ft 0 21/32 in	
Gap	7 ft 6 in		7 ft 0½ in	7 ft 6 in	7 ft 6½ in		
Dihedral					1°		
Height					15 ft 3in		
Empty Wt.	3,000 lbs			2,800 lbs	3,350 lbs	3,350 lbs	
Loaded Wt.	5,000 lbs				4,750 lbs	4,750 lbs	
Useful Load				≥1,500 lbs			
Duration				≥20 hrs			
Areas, Sq. Ft. (upper wing including ailerons)							
Upper Wing					551.30		
Lower Wing					301.83		
Ailerons					93.70		
Rudder		27			28.00		
Horizontal Stabiliser					63.60		
Vertical Stabiliser					24.10		
Elevators		48			49.00		
Non-skid Fins					25.00		
Surface				800 sq ft			
Fuel, US Gal.					200 (main)/ 12 (aux)		
Oil, US Gal.					8 each motor		

Notes on Sources:
1. *Flight*. Issue of 17.07.14.
2. Curtiss 1915 Catalogue, Original in Curtiss Museum, Hammondsport. Copy in RAFM J.M. Bruce Collection Box 22.
3. Curtiss advertisement Transatlantic Model Curtiss Flying Boat (with fins). Copy in Curtiss Museum, Hammondsport, USA.
4. Dimensions given by Porte in his letter of 14.08.14, to Sueter. Offer of 2 Curtiss Flying Boats. TNA T20/22/15.
5. Lt W. Capehart's letter to Bureau of Construction & Repair, 15.11.16. Propeller: Curtiss regular, Walnut 8 ft x 5 ft 6 in
6. Casey, L.S. *Curtiss – The Hammondsport Era 1907–1915*, Crown Publishers, NY, 1981, P.191.
7. Typed dimensions chart. RAFM J.M. Bruce Collection Box 22 Felixstowe File.

Price packed for shipment FOB Hammondsport, $25,000.
Engines: British H-4: 100-hp Curtiss OXX; 120-hp Beardmore; 100-hp and 125-hp Anazani; 150-hp Sunbeam Crusader.
Danish America: 90-hp Curtiss OXX-2.

Above: This boat with a single tail support boom may well be the Curtiss Cruiser. (via J. Gertler)

delivered at an additional total cost of $112,590.[45]

T.D. Hallam, a Canadian, had learned to fly on Curtiss flying boats at Hammondsport before the declaration of war. He eventually found his way into the RNAS and ended up at Felixstowe. He recounted his experiences in his classic book, *The Spider Web – The Romance of a Flying-Boat Flight in the First World War*. He recalled that *the early Curtiss H-4 flying boats were comic machines, weighing well under two tons; with two comic engines giving, when they functioned, 180 horse-power; and comic control, being nose heavy with engines on and tail heavy in a glide. And the stout lads who tried impossible feats in them usually had to be towed back by annoyed destroyers… the Navy people could not understand anything that could not be dropped with safety from a hundred feet, or seaworthy enough to ride out a gale.* [46]

This assessment is harsh. The Curtiss engines were changed for Anzani engines that, although rated lower, were more reliable and therefore considered more satisfactory. The machines proved to be excellent pilot trainers, it being possible to make every kind of bad alighting without damage to the hull *when operating in sheltered waters* (author's italics). Propeller shafts sometimes broke and it was not unusual to spend an hour changing plugs before a machine could get away. It was often only just possible to keep in the air with cylinders missing.

These boats have been known to climb as high as high as 6000 ft. but usually they were only just able to get off the water, and struggle up to 500 or 1000 ft. The least sea caused spray to be thrown over the pilots. It was a common sight to see clouds of steam generated from the sea water going over the engines. A certain number of patrols were carried by these machines early in the war from Felixstowe. The armament was 2 16 lb hand/bombs and a rifle. The operational success was not great, but these boats paved the way for the important developments which followed.[47]

Most of the pilots who made the *Large Americas* a success had spent hours training on these boats.

The shortcomings of these early boats were such that it would take a lot of work to improve the design to where it could operate under North Sea conditions, but progress was already underway under the hand of Porte. He had started to experiment with modifications to the hull in an effort to improve performance.

In early 1915 it was decided to establish a Naval Air station at Gibraltar and two "large seaplanes of the Atlantic type and 4 aeroplanes have left England. These machines will be supplemented when the station is erected."[48] These machines would have been H-4 boats 1236 and 1237 that were shipped to Gibraltar around May 1915. They performed patrols in the Mediterranean, America 1237 lasting until February 1916.

The first H-4 was written off at Felixstowe in January 1916. Declared obsolete in August 1918, three managed to survive the war. Serial 3580, which had been modified to become the Porte I, was the last *Small America* boat afloat. 3545 was scrapped in November 1918. The hull of 3548 had been sent to the Agricultural Hall, Islington, London, for preservation, but along with other priceless exhibits was later destroyed.

When the US entered the war on the Allies side, many personnel were trained in the UK and France. Some US Naval aviators were trained at Felixstowe and flew the *Small America* boats. 3587 was delivered to Felixstowe for erection on 16 March 1917. Ens Phillips Ward Page was flying this machine solo on 17 December that year. He had just performed a perfect alighting and taken off again. When he was at about 40 feet, for no apparent reason, he sideslipped to the left and nosedived into the water, the left wing apparently striking first. An F-boat had just alighted and taxied to the stricken machine that was floating with half the wings submerged and the tail vertical. Three motor boats from the Station and several life boats from trawlers anchored nearby also arrived. They were unable to right the aircraft or get to the pilot. On towing the wreck to the beach it was discovered that the upper part of the bow had been torn away and the pilot had been washed out. No trace of his body had been discovered by the following day.[49] The flying boat was a complete wreck and was deleted on 12 January.

White & Thompson Ltd acquired the exclusive rights for the sale of Curtiss aeroplanes, flying boats and engines for Great Britain in 1914. Porte was one of the company's pilots and flew the English Curtiss Model F flying boat that Glenn Curtiss brought to the UK in 1913. As such all negotiations for the supply of Curtiss boats should have gone through the company and this was the basis of the dispute that the company, renamed the Norman Thompson Flight Co, had with the Curtiss Company.

Norman Thompson built their own twin engined flying boat as the N.T.4 and N.T.4a, and to further complicate matters, they were also known as *America/Small America* flying boats. They differed considerably in design from the Curtiss America boats, being larger and their hull design owed little to the Curtiss boats. They are described in detail in Chapter 5.

The Felixstowe Line

At the end of January 1915 Porte was given charge of the Felixstowe Naval Air Station and in May he left Hendon altogether. He had also been appointed in charge of the Chingford and Chelmsford Naval Air Stations, but had to give them up at the same time as "Felixstowe became so important that it required my full attention."[50] At the time Felixstowe comprised 30 men and three small sheds engaged in local patrols with a few antiquated seaplanes. By the end of the war it had grown to some 1,000 men and a large number of buildings to be the largest Naval Air Station in England. Long range patrols were carried out with greatly improved flying boats. The station became an experimental as well as a war station under Porte's command.

According to Porte, a Capt Elder went to the US in March or April 1915 to arrange a very large order with the Curtiss Company. "The order was loosely placed and was badly placed, in fact, one could not make head for tail of it and it was over the difficulties of this contract and the frightful mess then made" that Porte went to the US to sort the contract out. The Admiralty had placed an order for 150 Curtiss boats. "I informed the authorities verbally that it was ridiculous – we could not do with 150 and there was nowhere to put them and so eventually in July or August 1915, I proposed I should go over to America to the Curtiss Company, stop the order at 50 and place an order for a further 50 to be of a new design of a larger type which I was designing myself. The 50 machines of the original type were delivered and I went over and designed the improved boats and ordered 50 of these machines."[51] Porte went to America in August and returned in September 1915.

Porte flew the early Curtiss boats and was aware of their shortcomings which included a lack of seaworthiness, poor hydroplaning characteristics and poor stability on the water. It appears that the H-4 boats were armed with bombs but definite confirmation and details are lacking. He obtained permission to experiment to produce a flying boat with the operational characteristics required by the RNAS. These experiments are fully covered in Chapter 2.

Porte also designed and built a large three engined flying boat at Felixstowe. Named the Porte Baby its history is detailed in Chapter 6.

By practical experimentation Porte was developing his knowledge of what was required for a successful flying boat. It was alleged by the Crown when Porte's contribution to the development of the flying boat was examined at the Royal Commission of Awards to Inventors after the war, that the development was *a natural progression* (author's italics) and Mrs Porte, Porte having died in the interim, did not warrant an award.

When the larger H-8/H-12 were received all the earlier Curtiss boats were referred to as *Small Americas*, the H-12 boats, F.2A , F,3 and F.5 boats becoming *Large Americas*.

The Curtiss H-7

The Curtiss Company were also developing the America boat under the direction of William L. Gilmore, Chief Engineer of the Company. What interaction there was with Porte is unknown. Unfortunately little is known about these aircraft, some only being known as they were recorded in photographs. The Curtiss H-7 was a three-seat, twin-engined tractor large flying boat known as the Super America. A cockpit for an observer was provided ahead of the pilot's cockpit where the two pilots sat side by side. It has been described as a transitional boat as it featured a hull with two outrigger booms bracing the engine beds to the horizontal stabilizers. The hull extensions were externally braced rather than being faired into "fins".[52] Two were ordered for Russia by July 1915. The boat was built at the Churchill Street plant in Buffalo. On its first flight on October 19, 1915,[53] the pilot, Theodore C. McCauley, who was flying solo, lifted off the lake surface the aircraft climbing at a steep angle. McCauley was alarmed and pushed the controls forward to avoid a stall. This had no effect, so he cut the power to the Curtiss V engines and the machine hit the water more or less under control. The overhanging panel on the top wing was damaged in the incident. The H-7 was tail heavy requiring 200 lbs of ballast in the forward cockpit to enable the aircraft to fly level. The machine was re-rigged, the engines and equipment

Above: Curtiss H-7. This photograph appears with the following caption in *Aerial Age*: "The H-7 Model Curtiss Flying Boat which made a flight lasting half an hour at Newport News last week with eight people aboard."

being moved forward.[54]

The test flights were concluded on December 13, McCauley having made 14 flights with the first and two with the second flying boat. The boats were then loaded to be shipped to Russia

It had been concluded that the hull needed redesign, and a new hull was constructed. This hull was used in trails conducted at Newport News, Virginia. These were long over water flights with full load and crew. The flights impressed the US Coast Guard and Curtiss donated a H-7 to the USCG.[55] McCauley established three new world records in a H-7 on 30 April 1916.[56] This would appear to be the machine with the new hull.

While on a flight to Washington, DC, one propeller shattered and the pieces flew into the other propeller while the machine was in a climbing attitude. At only 100 feet altitude, McCauley had little time to react as the machine turned upside down and crashed into the Potomac River. Two crew members were killed and McCauley was hospitalised due to a rigging wire striking him.

The problems that the H-7 boats had in Russia were due to the conditions they were shipped in and the fact that the Curtiss Model V engines were in the development stage. The aircraft suffered in the heat of the ship's hold and then the high humidity of the Russian seacoast. The propellers warped and vibrated causing problems with the engines to manifest themselves. The Russian ground crew were unused to handling such aircraft and their rough handling caused the bottoms of the hull to drop out when taking off and caving in on alighting.

Curtiss records show two H-7 boats built in April and four in June 1916. What happened to these boats is unknown but

Curtiss H-7 Specifications

Source	Note 1	Note 2
Length	39 ft 2 in	39 ft 1 in
Span upper	77 ft 0 in	77 ft 0 in
Span Lower	49 ft 2 in	49 ft 2 in
Chord	7 ft 21/32 in	
Gap	7 ft 5 ½ in	
Dihedral		1°
Sweepback		3° 27'
Net Weight	4,500 lbs	4,500 lbs
Gross Weight	6,500 lbs	6,500 lbs
Useful Load		2,000 lbs
Fuel (4 hours)		815 lbs
Oil Weight		145 lbs
Water Weight		110 lbs
Pilot Weight		160 lbs
Passenger Wt., ammunition or other load		760 lbs
Time to 2,700 ft		10 min

Source Notes: (1) "American Air Transport Services", *Flight*, 31.03.21, P.230.
(2) Curtiss Specification Sheet. Smithsonian NASM File AC-901659-01 Curtiss Model H-7. Casey uses the same figures.

Above: Another view of an H-7 with the modified hull with fins showing clearly the double tail booms..

parts are thought to have been used in manufacturing the H-8 and possibly the H-14. The H-14 was recorded by Lt (jg) W. Capehart, the USN officer at Curtiss, Buffalo, as having an H-7 hull with H-4 wings and control surfaces.[57]

The Curtiss Flying Cruiser

The English journal *Flight* published a photograph of a "Curtiss aero-yacht de luxe, built for the America Trans-Oceanic Co. It is luxuriantly fitted up and carries five passengers." This machine was apparently based on the late H-7. A single outrigger joined the rear centre section interplane strut to the leading edge of the tail plane rather than the two of the H-7. The machine was designed for the sportsman and the hull had accommodation for five. The two pilots sat in the large front cockpit that allowed for the accommodation of two passengers behind the pilots. Further aft was the cockpit for a mechanic.

Designed for sporting rather than for military use, the design and equipment of the Curtiss "cruiser", afford an indication of the trend that development undoubtedly will take once the war is over. It is a twin- motor flying boat of the pusher type with a speed range of 65-48 miles per hour and weighs over 2 tons fully loaded. The equipment includes a Sperry gyroscope, searchlight, marine running lights, lights at the top of each strut, and electric starters for the motors. Current for all these purposes is developed by two small fan-driven generators located on the inner end of the outrigger which supports the tail unit.[58]

Flat steel tube was used for the leading edge of the wings, the upper planes being in five sections. Spars were of "I" section spruce with ribs of pine, birch and spruce, except that the compression ribs are solid pine. Ailerons are on the upper wings only. Antiskid fins were affixed on the upper wing and braced the wing overhang. The lower wings were comprised of four sections. The two sections next to the fuselage were three feet long and were built rigid and covered with rubber matting in order that the mechanic could walk on them to service the engines. The outer wing sections were 19 feet in length.

The horizontal tailplane was carried on outriggers from the engine beds and struts arising from the stern of the hull. A large fin was mounted and an unbalanced rudder was mounted.

Two Curtiss 100-hp V-type motors were mounted between the wings. They drove opposite handed eight feet eight inch propellers and were equipped with electric self starters, though cranks were fitted for a hand start if required. The hull had an overall length of 34 ft 3 inches and a maximum beam of four feet at the leading edge of the wings. The hull was an ash frame and keel with mahogany and cedar sides covered with cloth.

Instruments included a compass attached to and synchronised with a drift indicator, and air speed indicator, fuel gauge, oil gauge, banking indictor, angle of incidence indicator, inclinometer. The throttle controls were placed amidships just between the two pilots.

Glenn Curtiss and Rodman Wanamaker formed the America Trans Oceanic Company in 1914 to promote the business that the success of the America flying boat was to bring to the aviation business. The outbreak of war ended the dream of a trans-Atlantic flight for the immediate future, however Wanamaker continued to negotiate with Curtiss for a trans-Atlantic aircraft and contracts exist of these proposals. In an interview with the New York World Curtiss noted that when

Above: This photograph has been identified as the small Curtiss H-8 pusher.

Above & Below: Although not positively identified, this flying boat with twin engines mounted in pusher configuration and a single tail boom, could be the Curtiss Cruiser or the American H-8. As with so many of these early Curtiss flying boats their correct identity is open to question.

the planed flight was abandoned he had purchased the America back from Wanamaker "on the understanding that he would build Mr. Wanamaker another machine with which to make the attempt as soon as circumstances would permit."[59] What operations were carried out by the Trans Oceanic Cruiser are not known.

J.L. Callan reported on the condition of the Trans-Oceanic Co's Cruiser directly to Glenn Curtiss as follows:

After having examined the Trans-Oceanic Co. Cruiser at Port Washington and talked to those who have it in charge I recommend that it shall not be flown until the hull now on the machine be replaced.

When this machine was first tried out at Hammondsport the motors were turning up 1650 R.P.M. and the boat was very light having been drying out in a barn at Hammondsport for over a year.

Unfortunately Callan did not date his report. It is possible that this report is post-war with the intention to bring the machine back into passenger service. Callan wrote that now

the Sperry equipment has been installed on the machine and the hull becomes heavier through water soaking, the machine does not handle well in flight and no reserve is to be had from the motors. While at first it was possible *to carry a dozen passengers* it is now hardly possible to get off the water with five.[60] (Author's italics).

This reference to a dozen passengers does not fit the description of the Cruiser in the 3 March 1917, article on the "Curtiss Flying Cruiser" in Aerial Age that stated that the Cruiser was with the America Trans-Oceanic Co at Port Washington, L.I.

The Company conducted training with a Curtiss F-Boat and offered the Navy a number of F-Boats after the US entered the war. The author has not been able to find any reference to the Cruiser being offered to the US Navy.

Post-war Trans Oceanic operated a Curtiss H-16 (the Big Fish), HS-2L, a Curtiss Seagull with a 150-hp Curtiss engine, and a Curtiss MF flying boat with a 90-hp Curtiss engine. The Company had contracts with the US Post Office and ran daily mails from their New York base to Chicago and Atlanta.[61]

The H-8 Pusher

This is another Curtiss flying boat that has a confused history. The flying boat referred to as the H-8 by Bowers in his history of Curtiss aircraft was a small biplane flying boat of pusher configuration. The change back to a pusher made it necessary to change from the twin booms of the H-7 to a single tail boom on the H-8 to avoid the propellers. This boom connected the engine section of the centre bay of the wings to the horizontal stabiliser at the intersection of the stabilizer and the fin. The hull appears to be similar (the same?) as that of the H-7. Only a single example of this aircraft was constructed in December 1915, at the Churchill Street plant.[62] No reason has been found for the reuse of the H-8 designation if indeed this was the correct designation.

Casey states that the pusher configuration was not desired by the British and the machine was changed to a tractor before being shipped to the UK, however the size of the pusher aircraft appears to be much smaller than the H-8 as supplied to

Curtiss Cruiser Specifications

Source	Note 1	Note 2
Length overall	29 ft 0 in	
Length Hull	34 ft 3 in	34 ft 3 in
Span upper	75 ft 10 in	75 ft 10 in
Span Lower	48 ft 0 in	48 ft 0 in
Chord	7 ft 0 in	
Aerofoil	Modified RAF 6	Modified RAF 6
Gap	7 ft 6 in	7 ft 6 in
Dihedral	1°	
Incidence	4°	
Span tailplane	16 ft 0 in	16 ft 0 in
Chord tailplane	6 ft 3 in	
Area, Sq. Ft.		
Wings upper	457.63	
Wings lower	301.83	
Ailerons	46.85 (each)	46.8 (each)
Non-skid fins	12.4 (each)	
Horizontal tailplane	63.9	63.9
Elevators	24.9 (each)	49.8
Fin	34.9	34.9
Rudder	31.2	31.2
Hull Weight (with seats, accessories, etc.)	885 lbs	885 lbs
Empty Wt		2,240.1 lbs
Engine Wt. (Curtiss OXX-2)		1,316.5 lbs
Fuel & oil Wt.		620 lbs
Net Weight	3,556	3,556 lbs
Useful load		1,100 lbs
Gross Weight	4,656	4,656 lbs
Fuel	90 gal	
Climb in 10 min	2,000 ft	760 lbs

Source Notes: (1). "The Curtiss Flying Cruiser," *Aerial Age*, 5 March 1917, P.718.
(2). Typed specification table. RAFM J.M. Bruce Collection, Box 22 Felixstowe File.

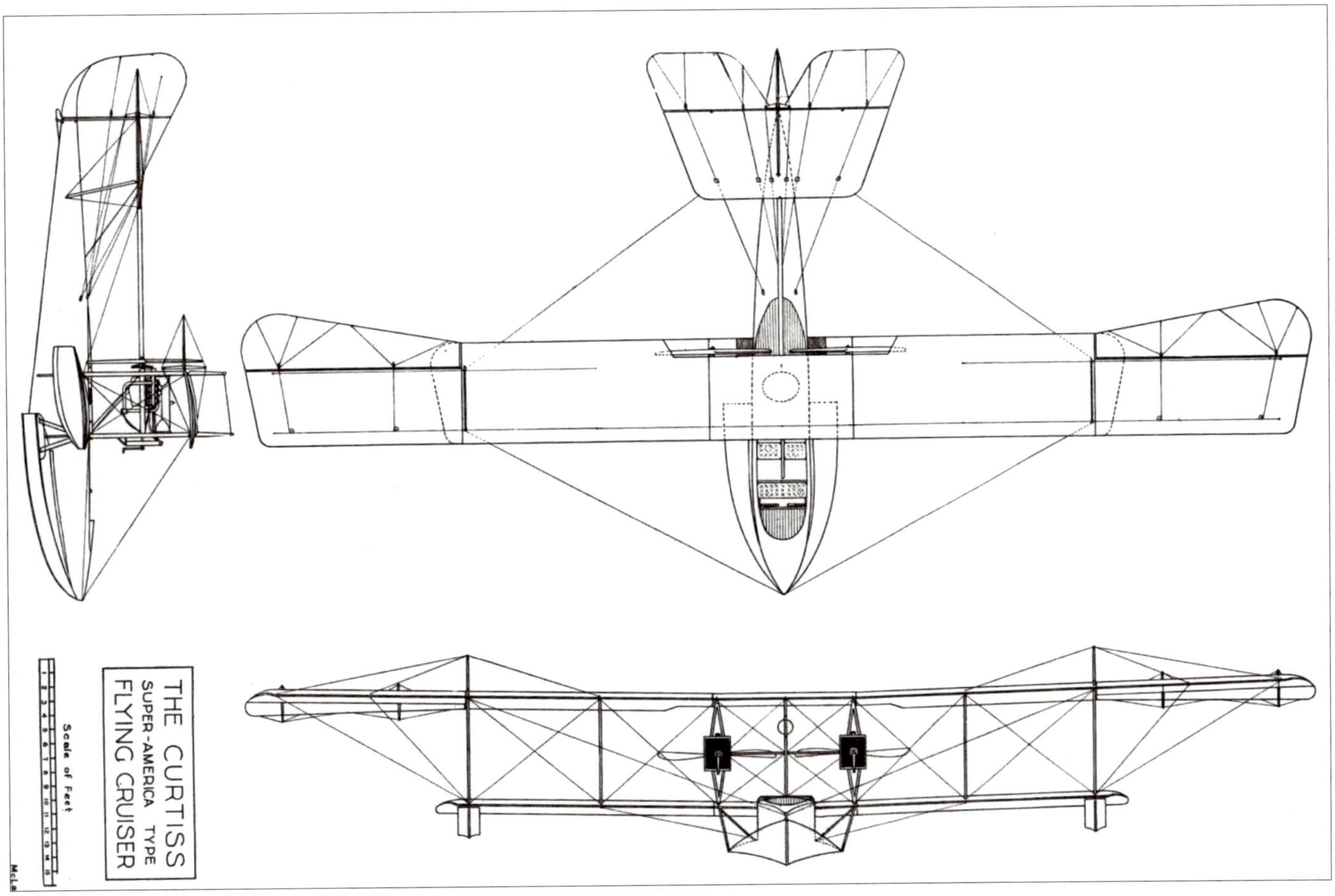

Above: Curtiss H-10.

the British. Without data it is impossible to tell and no data are available.

The H-10

The H-10 was a large twin engined boat with two booms connecting the engine nacelle to the horizontal tail. The power plants were Curtiss 160-hp VX engines. Aerial Age for 22 May 1916, reported that the H-10 had been wrecked near Alexandria, Va., on the 10th.

The accident was caused by the breaking of one propeller, pieces of which struck the other propeller, causing it to break also and thus making a quick forced landing necessary from a height of 100 feet. There was a strong wind blowing and one wing struck the water first, which, together with the speed at which the machine was moving, probably over 100 miles per hour, caused the craft to spin about and strike broadside causing its complete wreck.

There were five men aboard and none wore life jackets. Three were picked up by a tug which, assuming that this was the whole crew, moved away from the wreckage. Once one of the crew recovered enough to tell the captain that there were two missing, the tug returned but was not able to locate them.

Above: 8651 at Felixstowe. The pivot mounting for a machine gun on the bow indicates the presence of a gunner's cockpit. This machine appears to be in the Curtiss scheme of British Khaki upper surfaces. The colour of the lower surfaces of the hull is unknown.

Right: Front view of an H-12. The engines rotated in opposite directions.

Left: This may be 8654 or 8694, one of the original order for H-8 boats that were delivered as H-12 boats. The wings appear to have a dark dope to the upper surfaces. The hull could be English Khaki or grey while the fins and hull bottom has the black finish described in the appendix on camouflage and markings.

Above: The prototype HS-1 was developed from the unsuccessful twin-engine H-14.

Those rescued were Theodore Macaulay (*sic*), Phillip Utter, Mayo Dudley, newspaperman, while those that lost their lives were Louis Grant and Charles A. Good. None of the former were seriously injured.

The H-14

The H-14 was larger than the original America boat but smaller than the H-12. In November 1916, the Curtiss Aeroplane Co had written to the USN Bureau of Construction & Repair forwarding blue prints and specifications of the H-14. The machine was claimed to have an endurance of five hours "at rated horsepower" with a guaranteed high speed of 65 mph and a low speed of 48 mph. Climb to 2,000 feet was to be 10 minutes. "We feel confident that it can be relied upon for the above performance under all ordinary conditions and that it will do considerably better under usual normal conditions." The quoted cost was $22,500 FOB Buffalo.[63] Capehart forwarded the letter and documents to the Bureau of Construction & Repair and added the comments that it was understood "that this type boat is the type that the Company is bidding on in answer to the Army proposal for 144 water machines." He repeated the statement that it was the H-7 "boat body with H-4 wings and control surfaces. In addition there is a stay rod as shown running back to the horizontal stabilizer thus making the tail controls stronger than those in the H-4 type."[64]

The Navy considered the offer and noted that this "boat is similar to one recently constructed for Mr. Wanamaker and the weights, performance, etc., are taken from the actual construction." While the Company representatives stated that the Wanamaker boat had a high speed of 68 mph and climb to 2,400 feet (in ten minutes), they were not prepared to guarantee this. Notwithstanding these concerns, the Navy considered that two H-14 boats be purchased "by proprietary requisition for experimental purposes."[65] The hulls of the boats were to be stronger than that of the boat sold to the Trans Oceanic Co. A requisition for two H-14 boats, one equipped with Sperry Servo motor control was being prepared on December 6, the total revised price delivered to Pensacola being $35,664.

As noted above the H-14 was recorded by Lt (jg) W. Capehart, the USN officer at Curtiss, Buffalo, as having an H-7 hull with H-4 wings and control surfaces. By saying the machine had H-4 wings, etc., it is assumed that Capehart meant that the plan form was the same as the earlier boats. The H-14 had a very short overhang of the upper wing compared with the *America*. The wingspan of the H-series of boats varied according to the size and load of the designed machine but the same basic shape and layout was followed by Curtiss and Porte. The H-14 was powered by twin pusher Curtiss OXX-2 engines. Sixteen machines were ordered by the US Army before the prototype had flown (Army serials 396 – 411).

The USN representative at the Curtiss Aeroplane Co noted in August 1917 that the H-14 was "suitable for carrying Lewis gun...(and) ready for demonstration.[66] According to Robert Casari the Army had decided to cancel their "Twin-Hydros" by July 1917.[67] The Navy had ordered two H-14 boats (Bu Nos. A-145 to A-146). This contract was later cancelled and it is thought that the Navy's machines were never started as the Army agreed to allow the Curtiss Company to sell the 16 "twin motored military hydroairplanes" if after final inspection and official tests the machines "are found suited to Navy purposes."[68]

The performance did not meet specifications and the work on the machines was suspended. The prototype was converted into a single-engine boat and given the designation HS-1 (H boat, single engine). When fitted with the Liberty engine the machine became the successful HS-1L that carried the USN anti-submarine efforts from French bases after the US entered the war in 1917.

The H-16

The H-16 was the Felixstowe F.2A built by Curtiss and at the Naval Aircraft Factory (NAF). Despite being claimed as a Curtiss design, it was the Felixstowe boat modified to use Liberty engines and American construction methods.

The British Curtiss H-8 and H-12

Of the model numbers applied to the Curtiss H-series of flying boats up to the H-16, only seven can be confirmed as being

Above: A USN H-16 Bureau No. A-794. Reported as 100% complete in April 1918, A-794 was shipped overseas from New York on the Jason during the week ending 25 May. Assigned to Killingholme it was recorded at Killingholme on 13 August when it had to return from a convoy escort with engine trouble (Crew: Ens Ives; Ens Allen; CMM Lamont C. Fisher; E2c(r) Donald Phelps.). The machine was test flown throughout August with no patrols being recorded. As with so many of these flying boats it there is no record of its fate; not reported by any station; "Sent overseas not returned".

Below: An H-16 under construction at Curtiss Aeroplane and Motor Corp. Note the flap at the front of the aileron that folds back over the aileron gap when in flight. The fabric covered rear fuselage section is well illustrated. This would often break open in rough seas and was replaced by ply sheeting on many of the boats. The hull appears to be varnished on all wooden surfaces.

Above: Curtiss H-12 Bureau No. A-767 on the step, 24 November 1918. A-767 was reported as 100% complete in February 1918 and shipped from Buffalo to Pensacola that month. The Naval Air Station reported that it was receive d on 19 March with deficiencies. Liberty engines SE27/35 (port) and SE129/37 (starboard). On 9 March it was reported that an aileron was torn loose in a storm. During April balanced ailerons were trialled in moderately rough air conditions. In May Liberty engines 0993/212 and 1755/386 were installed. It was OOC during June awaiting parts. It was then used for instruction and on 24 July it was reported to have been the first to have heavy planking put on the step. A-767 had flown 58:10 hours and was still in excellent condition. In July it was with Pensacola Squadron VI and was used for training during July to November. The Squadron was now reported to be Squadron II and it is assumed that the ram insignia was associated with this squadron. On 29 November 1918, the boat was so "badly out of repair and adjustment that it will be prohibitive in time and expense to rebuild." It was RBS. Engines 0993/212 and 1755/386 were installed: TFT 381:50 hrs. The boat was officially stricken on 1 July 1919.

built. In the case of some of these machines, photographs are the only records available. The existence of two distinctly different machines as the Model H-8 only confuses the unravelling of the sequence of Curtiss manufacture even further. It is assumed that the H designation was applied only to twin engined flying boats based on the original Curtiss America. This assumption is given weight by the redesignation of the H-14 as the HS-1 when it was reconstructed with a single engine.

The H-8 sold to Britain was a large twin tractor machine without a tail boom. It was completely different to the other Curtiss H-8 and seemed to owe more to the H-7. This would have been the machine Porte claimed to have designed.[69]

> *In July or August 1915 I proposed I should go over to America to the Curtiss Company, stop that order (for the 150 Curtiss flying boats) at 50 and place an order for a further 50 to be of a new design of a larger type which I was designing myself. The 50 machines of the original type were delivered and I went over and designed the improved boats and ordered 50 of these machines of the larger type and in fact the last two machines of that 50 are coming over now and these are the machines which have proved so tremendously successful in the last few months. In fact they have revolutionised seaplanes and these were my design.*[70]

Curtiss material states that the H-8 was "the largest flying boat in the world" and that the

> *machine can be used for pleasure, or can be provided with a gun, and used for war purposes.*
>
> *The enormous capacity of this Flying Boat, over a ton useful load, is co-existent with the high speed of flight of 70 m.p.h. The gross weight with full load is 6850 lbs., or 6.2 lb. per sq. ft. of wing area. With provided fuel capacity, the cruising radius at full speed is 440 miles in 6.1 hours; in normal flight with the motors throttled part of the time, the duration is increased to 10¼ hours. Thus with the two extraordinary VX motors of 160 h.p. each, a highly desirable excess of power is provided which need not be used under average conditions.*
>
> *...The wide fins extending two feet beyond the beam lend*

Above: Two RNAS personnel pose with a dark coloured Curtiss H-4 boat in the background.

excellent stability; and the pontoons safeguard the machine against any considerable rolling in the water.

...The upper wing has an overhang of 12'-3" at each end for efficiency and structural reasons, wherefore 59% of the total 1110 sq. ft. wing area is advantageously located in the upper wing. Nine separate panels compose the prodigious wing spread, three integral with the boat assembly, the others demountable for shipment and storage. The internal wing construction is not only light and rigid, but also exceptionally sturdy, employing beams and ribs of I-beam section cross braced with nickeled steel piano wire....The nose portion of the wings are covered with 2-ply birch sheathing in order to preserve the exact curvature by preventing sagging between the ribs....That part of the lower wings within two feet of the body on each side is also sheathed in wood to serve as "sidewalks."

...The power plants consist of two Curtiss Vx Model 160 H.P., 8 cylinder V Motors, mounted between the wings, 11 feet apart, centre to centre and enclosed in streamline hoods...(there was) *one gravity tank behind each engine...being supplied by windmill operated pumps, or by a hand pump in the hull.*

The hull is of the latest Curtiss Type. By means of the Fins, an exclusive feature of the Curtiss Boats, is secured a larger planning surface and flotation stability, without greatly increased bulk and head resistance. The planning bottom is "V" shaped, with the proper concavity to deflect water to each side rather than upwards. The nose has a decided overhang with consequent advantage to the fore and aft stability in the water, and with added planning area....

The seats are side by side; enclosed in a mahogany finish cabin; with a view ahead through the celluloid cabin windows unobstructed by the wings. Seats are formed of sheet aluminium, have leather sliding adjustments fore and aft, and are comfortably upholstered in leather. The cushions are stuffed with Kapak to serve as buoyant life preservers.[71]

The hull was typical Curtiss construction with two 3/32" layers of mahogany below the water line, laid across each other with a layer of Sea Island Cotton fabric glued between the two layers. The planking above the waterline was 7/32" cedar. The hull was covered with light coats of high grade copal varnish. The wing tip pontoons were of 1/8' ply planking. It will be noted that Curtiss was offering the H-8 to sportsmen although no record of a sale to a private individual has been found to date.

50 of these large boats were ordered by the British (8650-8699). The first of these Large America boats, 8650, was received at Felixstowe in March 1916.

When 8650 arrived it was discovered that the hull was very elaborately fitted up and the first thing done was to remove a great deal of unnecessary woodwork. The Curtiss engines were unsuccessful chiefly due to a bad attachment of the airscrew boss, the airscrews failing for the same reason. One went through the shed roof, and another burst in the air. On one occasion pieces came through the hull behind the pilot and

Above: The first and only British H-8 with Curtiss engines at Felixstowe. Although photographic confirmation is lacking there does not appear to be an opening for the front gunner. The H-12 supplied to the USN did not have a gunner's position at the bow.

almost cut the boat in half. Maj Cooper, who was piloting at the time, had enough control wire remaining to bring the machine down to a successful alighting.

Curtiss 160-hp VX engines were originally fitted to the H-8 but they could not get the boat airborne with full combat load. They were replaced by 240-hp Rolls-Royce Eagle engines and take off was finally achieved. Fitting Rolls Royce engines in place of the Curtiss engines reduced the horsepower load, but the wing load was much higher than the Small America and the hull submerged deeper with a pronounced lack of buoyancy forward. As a result the "hump", that occurred at about 18 knots, was very pronounced. The hump speed is that where water resistance reaches its maximum value.

In the air the machine had a better performance than any other seaplane to date, reaching 76 mph and a ceiling of 11,000 feet. The defect of the type was insufficient buoyancy, but it had a good range and established the parameters for the right type to do effective work over the North Sea.[72] The H-8 carried out three operational patrols in July and August.

The bottom before the step was made a double curve and was extremely efficient as a hydroplane. More buoyancy was provided by making the fins of great thickness, and by giving them steeply arched tops. The hull structure was considerably lightened and a very good scheme of transverse bracing of steel tubes and flanged plate work was introduced. Unfortunately insufficient strength had been provided in the vertical direction along the tail. This was remedied after trial and subsequently gave little trouble.

Above: This poor photograph is included as the original caption states that it was an H-8 at Felixstowe. The radiators do not appear to be the usual circular Curtiss shape and it may well show the H-8 as delivered to the British.

The tail was perfectly flat beneath and arched above and with rounded corners. There was no external keel and the keelson was shallow, the transverse floors were of spruce, and had big lightening holes cutting across the grain.

The Curtiss engines were not accepted and new mountings had to be made for the Rolls Royce Eagle engines. The petrol pipes on 8650 were without expansion joints and were connected by screwed unions which depended on shellack to make them tight. The petrol pumps had to be replaced by ones made on the Station, and the gravity tanks built and fitted into the centre section of the top plane. The results of these trials and modifications must have gone back to Curtiss in the USA as the next machine to arrive at Felixstowe in October 1916, 8651, had a larger hull. Considering the time between when the first and second machines were accepted at Felixstowe, it is assumed that the relevant drawings would have been prepared and shipped to Curtiss. To date no correspondence has been found that can provide light on the exchange of information between the British and the Curtiss Company. As with so much of the America story there are still many areas of uncertainty. The machines were now called H-12 *Large Americas*, the earlier boats being termed *Small Americas*. Curtiss also mentioned an H-9 being supplied to the British but this flying boat has proved elusive and has not been identified to date.[73]

The rest of the boats under the contract were delivered without engines at a price of $16,200 each with propellers, and $15,930 each without propellers. By replacing the Curtiss engines with 250-hp Rolls Royce Eagles the loading was reduced from 27 to 18 lb/HP.[74] The Large Americas were eagerly awaited at operational stations and they soon proved their worth.[75]

The results of testing and operating the new boats by the British were disappointing in some respects. Their hull was structurally weak. Capt A.E. Bolton recorded the disadvantageous of the H-12 in his post-war recollections. (See Appendix No.). Capt David Nicolson considered that the H-12 "was probably the worse example of boat building that could be imagined, it having no less than four consecutive planks butted – not even scarphed – on the same timber, which had a siding of only 3/8 inch, the line of the butts being in line with the step, where the boat was naturally weakest."[76] Therefore 8650 became the candidate for a new Porte hull. With this hull it was known as the Porte II and was to prove successful. It performed a number of patrols before being sunk on 30 September 1916. The development of the Felixstowe boats from the Curtiss boats is detailed in Chapter 2.

According to Maj Vernon the H-12 was followed by the H-12A, the name given to the improved H-12.[77] The chief change between the two being a larger tank capacity and better accommodation for the observer in the bow with provision for a gun ring. They also had a revised cockpit canopy. USN documents indicate that the H-12B had a considerably larger horizontal stabiliser and elevators. The USN version with the Liberty motor was the H-12B.[78] It has also been reported that some British H-12 boats had a wider centre section in order to fit Sunbeam Cossack engines. According to the Appendix XLI of the "War in the Air", the H-12 with Porte hull was known as the "H.12 Converted Large America."

Early boats had a pillar mount for the bow cockpit machine gun. Later a Scarff ring was fitted with single or twin Lewis Guns and a revised cockpit enclosure. Changes were built into the boats as they were manufactured and after delivery. With the ability to carry four 100-lb bombs, or two 230-lb bombs, an operational ceiling of 11,00 feet and a speed of 85 mph, the H-12 offered the British a boat that was vastly superior to any previous model. Notwithstanding the problems that soon were evident, including a lack of defensive fire to the rear and an inherent structural weakness in the design in that it tended to be damaged on takeoff, the boats soon proved their merits.

With an endurance of eight hours, the H-12 could make scouting flights half way across the North Sea. They could operate in rough water that would prevent the floatplanes that equipped the RNAS from operating. The H-12 could also reach the southernmost position that the Zeppelins patrolled.

The size of these boats meant that new lessons had to be learned. Getting the boats on their trolleys was never an easy task except in a calm sea, and the tails of these boats were so weak that the least touch of a hard object punctured them. While experience was being gained a number of hulls were damaged by fouling their trolleys on the slipway.

The phrase "nothing further was seen" is continually repeated in the reports of marine aircraft on their long, lonely patrols. These two patrols were not the norm in that they sighted a submarine even if they could not carry out an attack. 8677 was to be lost on 24 April 1918, Claimed by *Oblt. RMS* Christianson. 8689 suffered a similar fate being shot down on 4 June 1918. The crew were interned in the Netherlands, their boat becoming L-1 in the Dutch Navy.

As a school machine the H-12 boats were not good, the tails being stove in or damaged in one way or another by ham fisted students. Nevertheless a great deal of training was carried out with these boats.

The planning bottom, besides being efficient, was stiff compared with the rest of the hull structure. The bows would meet the seas with very hard blows, and cases occurred where the body of the boat could not stand up to the sudden loads. When this occurred the structure failed in compression at the sides and top of boat forward of the wing roots; yet at the same time, some hulls stood up extremely well to adverse conditions.

Apart from the local weakness, the tails developed a curious general flabbiness, particularly those boats which had been down at sea and towed home. It showed itself by unexpected variations in nose or tail heaviness. A machine would start out all aright and after a short while in the air, suddenly be found to be out of balance.

Some of the tails developed a twist, and as there was no method of trueing them up, the tail plane stays had to be made longer or shorter, to get the (tail) *plane horizontal. These changes alas, were not permanent.*

One point in favour of these hulls was the bulkheading in the tail. On more than one occasion hulls damaged on landing at sea, were kept afloat for very long periods, enabling the crews to be rescued. The most striking case, was that of a leaking machine, being kept afloat in the North Sea for four days.[79]

Out of a total of 360 completed three to four hour anti-submarine patrols from Felixstowe, only 11 forced alightings took place, and several of these were only through such defects

Above: H-12 boats under construction in the Curtiss Buffalo, NY, plant. The cockpit enclosure identifies these as H-12B boats. The hull to the left has the uncovered wings attached to ensure the fit of all parts, while the hull to the right has the uncovered tailplane attached. Covered wings are laid out at the far right. This workspace is less cluttered than most photographs of the Curtiss factories.

as oil or water pipe joints failing. Maj Vernon recorded that it was "really the Rolls Royce engines that saved the situation." [80] Even Rolls Royce engines could have their problems as flying boat pilots at Killingholme recorded that the reduction gear would shear right off some of their Eagle motors.[81]

Capt A.E. Boulton recalled the introduction of the H-12 and how it changed the situation in the North Sea. His full reminiscences are given in Appendix No. They were an immediate success in anti-submarine work and against Zeppelins. He also was aware of their deficiencies and considered that the F.2A as developed by Porte overcame the majority of these.

The H-12, along with the Porte Baby, was declared obsolete at the Armistice, the Felixstowe boats taking over the roles pioneered by these earlier boats. *Small America* 3549 was shipped in September 1918 to the Agriculture Hall, Islington, "for storage for exhibition after the war," arriving on the 23rd minus planes & engines.[82] Unfortunately none of these boats were preserved.

The USN and the Curtiss H-12

The USN was interested in the Curtiss boats developed for the British and ordered a single example, Bureau No. A-152. This machine did not have a position for the gunner/observer in the bow, and it is thought that this machine was probably in the same configuration as the H-8 delivered to the British although only references to the designation H-12 has been found in official documents.

A-152 had a long development time as it was delivered to Pensacola in March 1917 with a Curtiss Company crew judging from photographs taken at the time. On 10 May it was reported as having its Curtiss engines fail while on trial. At the end of July it was still awaiting trials. During the following month it was reported as having dual control installed and by the 30th the repairs and alterations were practically complete. Lt B.L. Leighton flew a test in A-152 while the machine was being readied for its acceptance trials. In addition to Ens Cheney and Mr Schaeffer, a Curtiss representative, Mr King, was aboard. The flight of 30 October ended in a crash from which the passengers survived but only the two engines and self starter was salvaged. Leighton testified that the machine was unairworthy but he was blamed for the crash.

In September the British had agreed to release to the USN from the H-12 boats under construction by Curtiss boats No. 10, 11 and 12. The USN H-12 boats were used mainly for training although it was found that the smaller Curtiss F-Boats were more efficient and did not suffer as much damage from students due to their better construction.

Later *Large America* Boats.

Porte worked with the H-12 to develop a new hull and this emerged as the Felixstowe F.2A, probably the best flying boat of the war. This was further developed into the F.3 and F.5. The Felixstowe F.2A and F.3 boats were all known as *Large America* boats. The last Curtiss-built *Large America*, the H-16, was the F.2A modified to take the Liberty engine. Being Felixstowe designs they are beyond the scope of this chapter but the importance of the Curtiss flying boats to the British anti-submarine effort cannot be under estimated.[83]

A secret memorandum of February 1917 on future policy for the RNAS noted that the Porte Baby flying boat was not being proceeded with, and the *Large America* Type would be the patrol boat for the future. Unlike the Porte Baby flying boats that could only be accommodated at Felixstowe and Killingholme it was stated that although the *Large America* boats required fairly large sheds and slipways, they could

Felixstowe Anti-Submarine Patrols

Time	Compass Course	Course to be made good	Distance in miles	Time mins	Notes
13:42	106°	100°	70	70	Shipwash Port Beam 1:55 p.m.
14:30					Sighted enemy Sub: 4 mls S'board bow steering NNW. Approx position 10 mls ESE Nth Hinder Circled around spot for 15 mins
14:45					Resumed course
14:49					Picked up 4 destroyers & signalled information re position of sub.
15:14	270°	280°	73	83	Nth Hinder 3 mls port beam 3:32 pm. Shipwash 4:15
16:37					Landed Felixstowe

Curtiss H-12 8677.
01.11.1917.
Pilots: Flt Lt Ball; Flt Lt Faux.
Engineer: Hollidge.
W/T Operator: Grey.
Start: 1:40 pm. Landed: 4:37 pm. Time: 2 hrs 57 min.
Distance travelled: 192 mls.

Remarks: On sighting the Enemy Submarine altered course so as to gain position between submarine and the sun, and attack unseen. Were observed by submarine which submerged immediately, about two mins after out first sighting her. Circled round vicinity for 15 mins, but nothing further was seen.
Dropped two smoke bombs on return journey.

Felixstowe Anti-Submarine Patrols

Time	Compass Course	Course to be made good	Distance in miles	Time mins	Notes
12:05	97°	94°	52	45	12:50 N Hinder abeam.
12:50	140°	135°	10	10	
1:00	Courses various		10	10	Sighted conning tower of submarine on port bow 4 – 5 mls away. Altered course slightly, but submarine submerged about 3 mins before we could reach it.
1:10	150°	145°	08	08	
1:18					Sighted 6 mine sweepers on port bow.
1:18-2:45	Courses various				Circling around mine sweepers.
1:28					Sighted 4 destroyers & one kite balloon.
2:45	275°	280°	26	30	3:15 Over N Hinder
3:15	268°	274°	52	57	3:51 Shipwash 1 ml starboard beam.
4:12					Landed Felixstowe.

Curtiss H-12 8689.
Date: 31.10.1917.
Pilots: Flt S/Lt Young; Flt S/Lt Barker.
Engineer: Clarke.
W/T Operator: Norris.
Start: 12:05 pm. Landed: 4:12 pm. Time: 4 hrs 7 min.
Distance travelled: 240 mls.

Source for both: "Felixstowe Seaplane Base Anti-submarine seaplane Patrols." TNA AIR1/639/17/122/193.

Remarks: Visibility out, fair. Visibility return, very bad, a thick haze.
When the submarine became visible through the haze, she was distant 4 – 5 miles and was rapidly submerging, and disappeared from sight within half a minute after being sighted. We circled around the position, but nothing further was seen, no time and position were sent to base by wireless. Immediately on getting in touch with mine sweepers and destroyers information re submarine was sent by Aldis lamp and acknowledged. At 2:45 pm we signalled our intention to return to base to destroyers.

Above: A-152 was the first H-12 ordered by the USN. Delivered to Pensacola in March 1917 it was reportedly taken from a British order. It is considered that this seaplane represents the configuration of the first H-8 delivered to Felixstowe although US documents refer to is as an H-12.

Below: Close-up of the same crew. This seaplane has the early cockpit canopy but no bow position. Photograph dated 7 April 1917. Testing was delayed with engine trouble and when tested by the USN it crashed.

be accommodated at the following stations: South Shields, Killingholme, Yarmouth, Grain, Calshot, and Felixstowe.[84]

Post-War Claims with Respect to Curtiss Flying Boats.[85]

Rodman Wanamaker made a claim against the Aircraft Manufacturing Company (AMC) for royalties for flying boats produced by that company during the war.[86] The papers for the case reveal that a Mr William E. Wood, acting for Wanamaker, entered into an agreement with the AMC on 13 November 1914, whereby the AMC agreed to prepare and complete plans and specifications for the construction and completion of a seaplane in accordance with the ideas and schemes of Wanamaker, and that such plans and specifications would be the sole property of Wanamaker. Any seaplanes built by the AMC using these plans and specifications built in the next five (5) years would entitle Wanamaker to a royalty of 10%. A second agreement of January 17, 1917, was along similar lines.

The AMC stated that it did utilise these plans and specifications but solely in connection with the construction of "certain craft we were ordered by and supplied to" the Royal Navy. A Minute in the file notes that "it may be suggested that these designs in fact were the production of Colonel Porte whilst he was in the Service, and that Wanamaker got them from him." In any case it was clear that Wanamaker was not acting as an agent for, or with authority from, the Curtiss Company.

The history of the America boats as related in the files concerning these claims is at odds with actual events. The file states that Porte brought over the Curtiss flying boat and claimed the design as his own. He sold the *America* boat to the Admiralty, then altered the design and got the AMC to build boats to this design, and that he convinced Wanamaker to finance the building of the "second boat".

Porte was instrumental in the RNAS obtaining the Curtiss *America* and its sister ship, but as has been related, the order to purchase came through official channels and Porte was asked to carry out the negotiations. While Porte did claim in his application to the Royal Commission on Awards to Inventors. that he had designed the *America* flying boat and the Curtiss H-8/H-12, in truth he probably only had input into both these designs, Curtiss engineers and Glenn Curtiss himself made the changes that were to improve the original America and made it a success. These changes, such as the fins, were the subject of patent applications by the Curtiss Co. Likewise, Porte's claim to have designed the H-8 is probably an overstatement. He no doubt produced the specification that led to the H-8 and his work with the first one delivered to Felixstowe led to the successful H-12. It is known that Porte worked with his American contemporaries when the US entered the war, and the Curtiss engineers also worked closely with the USN's Bureau of Construction and Repair. There was a constant sharing of information; the Curtiss Company sharing information with Porte and his Felixstowe team before the US entry into the war.

The statement that Porte had convinced Wanamaker to finance the building of "this second boat" is not clear, but appears to be in relation to the design that the AMC built. If the boats were ordered by the RNAS why would Wanamaker have to finance the building of the machines? What evidence was produced that led to the statement that Porte had Wanamaker finance the building of "the second boat"? Was the relationship between Porte and Wanamaker in carrying out this contract a means of avoiding Curtiss' patents? Given Wanamaker's continuing relationship with Curtiss and his further contracts with Curtiss to build a trans-Atlantic flying boat, it is not clear why he would be involved in a scheme to get around Curtiss's patents.

The hulls of the "America" boats ordered from the AMC were not completely replicas of the Curtiss design. The hulls were built by Saunders and were of different construction to the Curtiss design. The Crown acknowledged in the file notes that the boats built by the AMC were possibly the basic Curtiss design improved by Porte "when in service of (the) Crown." The contract with the AMC allowed for the Admiralty to pay "all reasonable costs" plus a profit margin. Therefore if Wanamaker's claim against the AMC was held to be valid, then the Crown would most likely be liable for the amount. These boats were already included in the Curtiss claim and the whole issue was very complicated. It was further complicated by a claim by the Norman Thompson Co that covered some of the aircraft in the Curtiss claim, as well as Mrs Porte's claim.

The Crown was unable to state who was entitled to these designs, whether Wood (and Wanamaker) or Curtiss, or Porte, until the action was tried and the alterations by each of them were worked out.

The reason for the second agreement between Wanamaker and the AMC is not known.

Claim by the Curtiss Company Against the British Government.[87]

Curtiss claim was that information as to the construction and design of "hydroaircraft" was furnished almost continuously throughout the war by Mr Curtiss personally or by officers of the Curtiss Aeroplane and Motor Corp to representatives of the British Government by their request. It was suggested that the matter could be resolved by referring it to the Royal Commission on Awards to Inventors and this was done, a 10% royalty being asked for as recommended by the Admiralty when an informal request was made in 1915, but Curtiss did not pursue a claim at the time as he did not want to embarrass the British Government while hostilities continued.

Patent matters "in greatly amplified and developed form, as exemplified by the America were communicated to the British Government by the delivery of the America and subsequently by the sister ship to the America. These novel features were communicated to the British Government by:

(a) delivery, to Charles Sumner and C.R. Bennett, New York agents for the Cunard Line, acting as agents for the British Consul General, the America to the *SS Mauretania*, and

Hull Comparison of RNAS Flying Boats

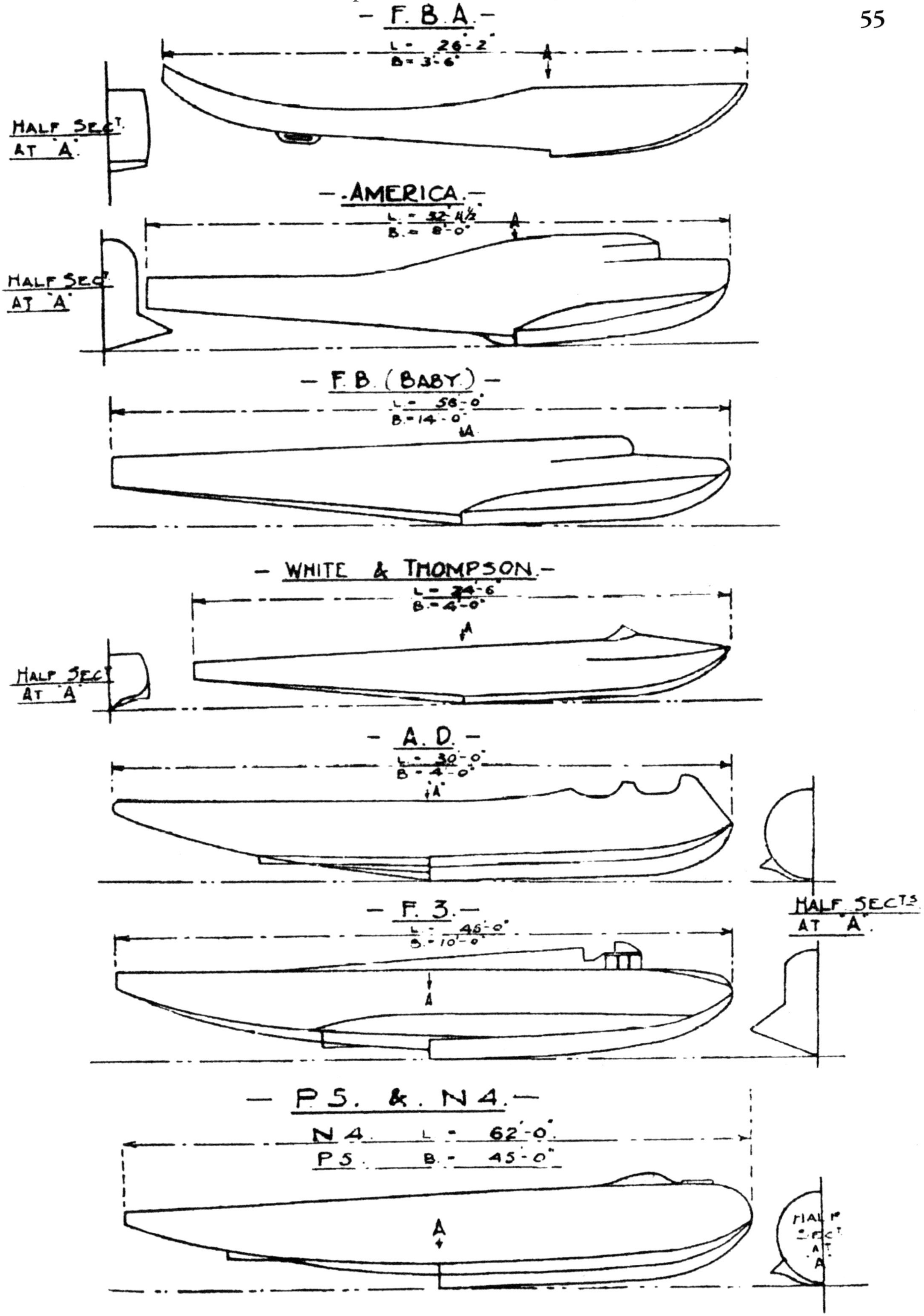

Above: 8660 displays the position of the front gunner and the early style of cockpit canopy. Note the way the lower hull surfaces colour comes up to near midway on the rear of the hull. Upper surfaces were probably Curtiss British Khaki. No indication of the lower colours has been discovered to date. On 29 October 1917, 8660 left Yarmouth with a crew under Flt Lt Leckie and Flt/Sub-Lt Bolton on a special patrol to search for a missing seaplane and mines near Brown Ridge. 10 Miles EbyS of Smiths Knoll conning tower of a submarine broke surface immediately below aeroplane. On turning to bomb the submarine had disappeared and could not be found. The periscope was sighted later and the boat increased speed increased to 90 knots, two 100-lb bombs dropped on the spot but no visible results were observed.

the "sister ship" of the "America" in a similar way to the *SS Lustinia.*

(b) Verbal communication by Lyman J. Seely, the Curtiss representative in the UK, to Cmdr Murray Sueter, Lt Spencer Grey and other officers in charge of the RNAS, in 1914.

(c) In January 1916 by verbal and sketch drawings of the Curtiss flying boat and developments by Seely to Cmdr Sueter.

(d) In the early months of 1915 by the delivery of four flying boats to British agents in New York to order dated 8 December 1914, for same.

(e) On or about 7 September 1915, by verbal description and demonstration of the Curtiss flying boat and method of manufacture given to Porte by Glenn Curtiss and members of his company at Hammondsport, NY.

(f) On or about 6 December 1915, by Seely to a Mr A.N. Smallwood with reference to "the stub wing type of construction and the 'V' shape strut and engine support", and previously to Porte in the USA.

(g) From October 1914 until the spring of 1918 the Curtiss company delivered to the British Government representatives at New York, Curtiss flying boats involving novel principles...These deliveries were practically continuous during the entire period and therefore reference is not made to individual shipment. The types shipped in chronological order were designated: Americas, H2, H4, H8, H9[88], H12 and H16."

(h) Along with each flying boat a set of plans and photographs was supplied to the New York agents, the 1st Civil Lord of the Admiralty, in London; to the Contract Section of the Admiralty; to the Officer in Charge of the Supply Depot, RNAS, White City; and one set with each boat.

(i) Information was supplied to British Officers in the USA including:

Lt Spencer Grey in the latter part of 1914 and early part of 1915.

Porte between 25 August and 6 September 1915, and when he was in the US for a short time in 1915.

In 1915 to Capt W.H. Elder, RN, who was stationed in New York City.

From 1915–1917 to Cmdr Maurice Edgerton, who was stationed in New York City.

From 1915 to the spring of 1917 the British Government

maintained officers attached to the Curtiss factories at Buffalo and Hammondsport. The greater part of this time the officer in charge was Lt. Cmdr. B.O. Jenkins, RNR.

(j). Mr C.M. Manly, the chief engineer for the Curtiss Co, proceeded to the UK to discuss hydro-aircraft matters, and remained there from 24 June to 15 September 1916. He made full disclosures to Sueter and officers of the Admiralty and to officers in charge of Felixstowe.

(k). The following Curtiss representatives were sent to the UK: James Tealy who took blueprints of the Curtiss Type flying boat to Felixstowe on 5 March 1916.

George L. Robertson, a Curtiss engineer, who remained in the UK from June 1915 to March 1917. He made full disclosure and made periodic trips to Felixstowe.

Clayton Roloson, a Curtiss engineer, was in the UK from November 1915 to July 1916, and spent most of his time at Felixstowe where he made full disclosures relative to the "so-called "T" type flying boat to Porte and other officers.

Curtiss' claim was for the methods of construction and manufacture as covered by his patents. The following British patents were tendered as exhibits:- 15,786; 113,965; 102,620; 113,964; 21,536; 12,648; 15,787 and 1102,261.

These covered the "novel features" that formed the basis of the claim, namely:-

(a) The boat hull incorporating a step "thus obtaining the longitudinal control necessary to permit the boat to leave the water and take the air."

(b) The side fins that increased the hydroplaning surface without adding greatly to the weight or wind resistance.

(c) The stub wing construction that permitted the building of portion of the wing surfaces as an integral part of the rest of the aeroplane structure.

(d) "V" shaped strut arrangement forming at the same time an engine bed.

The British refuted the claims of the Curtiss Company. While acknowledging the remarkable liaison between the Government and the Curtiss Company, the British Government noted that from about November 1914 onwards, constant research was carried out at Felixstowe for the purpose of improving the *Small America* flying boats purchased from Curtiss. The whole of the experience, knowledge and designs thereby acquired and perfected by His Majesty's Government were freely communicated to the Curtiss Company enabling them to build these flying boats for Britain. It was pointed out that the Curtiss H.16 was built from plans supplied from the UK, being the F.2A, while the F.5 (F-5-L in US service) was another Felixstowe design that was to be built by Curtiss, the Naval Aircraft Factory and by the Canadian Aeroplane Co. Approximately $11 million was spent in purchasing aeronautical material from the USA; a great proportion of which was spent on naval aeronautical materials. The Commission did not agree with the Curtiss claim and did not recommend that any payment be made to the company.

From RNAS Rigging Notes for America Flying Boat

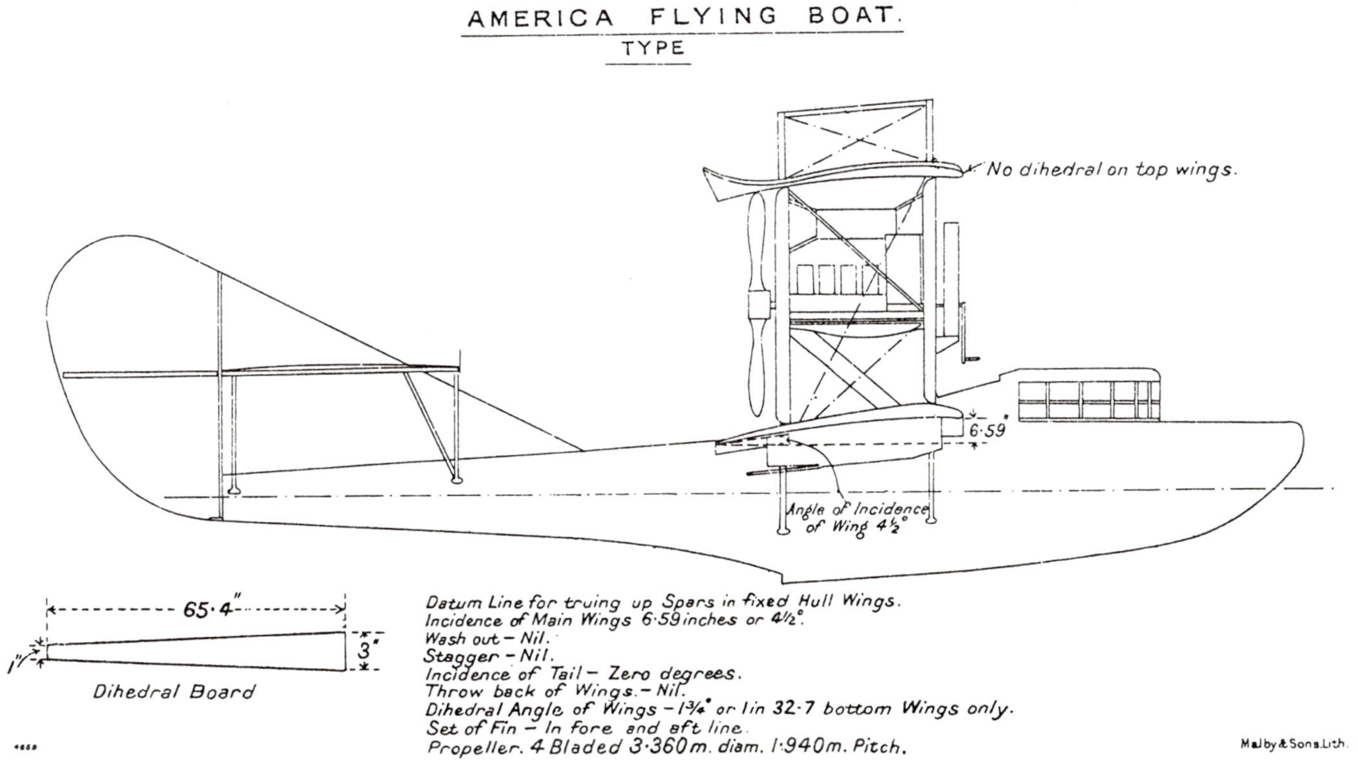

Above, Below, & Facing Page, Top: Construction of the original *America's* hull. The methods used were the same as applied to boat building. The thin planks may be seen laid over the substantial internal framework.

Above: The crowded conditions in the Curtiss workshop at Hammondsport show clearly why the machine had to be erected in the open.

Above: Close-up view of the engine mounts and upper centre section struts.

Above, Below, & Facing Page, Bottom: The original side-walled *America* hull.

America Patent Drawings, Following Five Pages

This is one of many patents that Curtiss took out covering the America flying boats. It is not known which model the drawings refer to but the balanced rudder suggests that it is an H-4 or later variation of the original model. Notwithstanding which model it is, the drawings give a good indication of the construction of the early Curtiss America flying boats.

G. H. CURTISS.
FLYING BOAT.
APPLICATION FILED DEC. 6, 1915.

1,246,016. Patented Nov. 6, 1917.
6 SHEETS—SHEET 2.

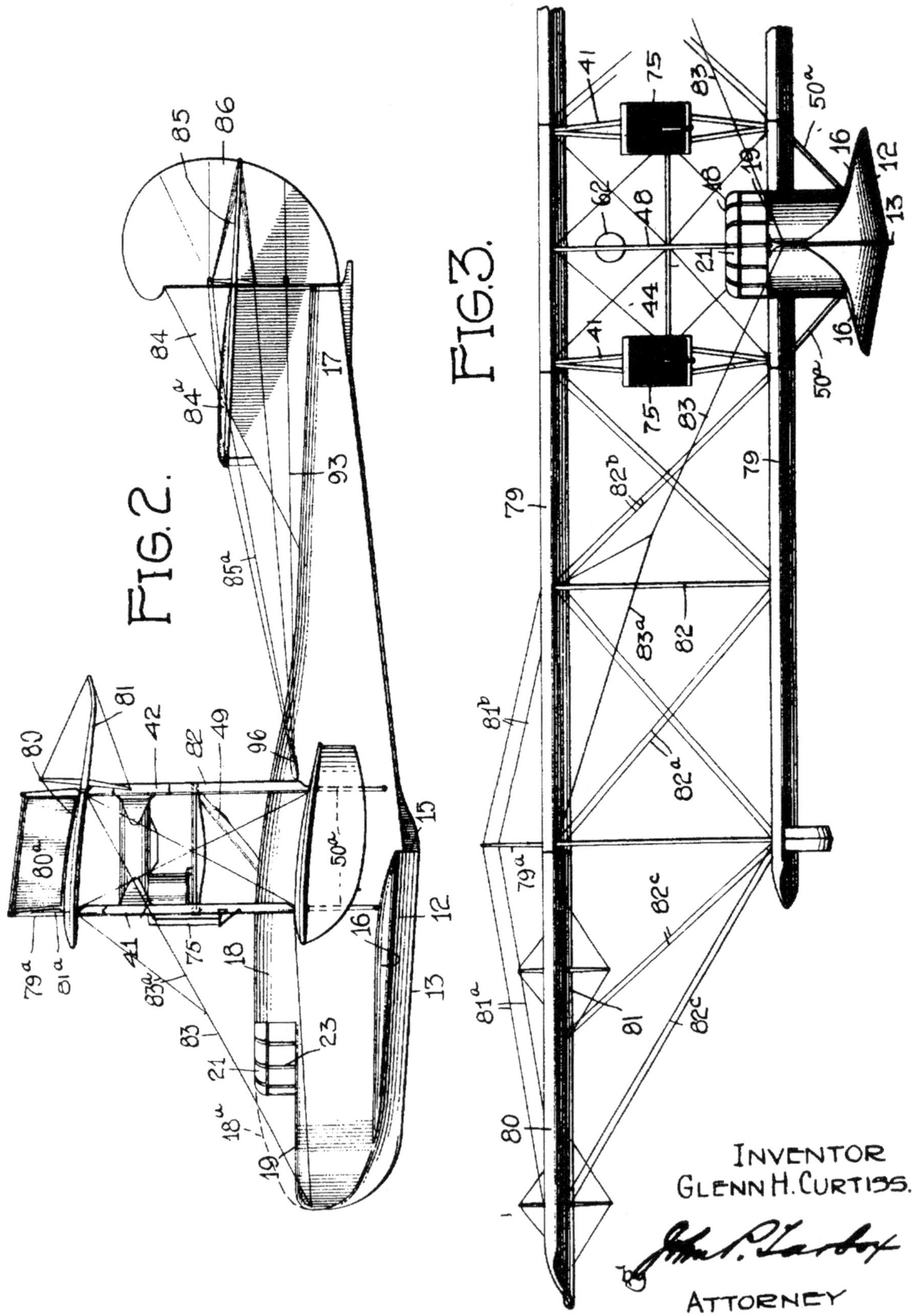

INVENTOR
GLENN H. CURTISS.

[signature]
ATTORNEY

G. H. CURTISS.
FLYING BOAT.
APPLICATION FILED DEC. 6, 1915.

1,246,016.

Patented Nov. 6, 1917.
6 SHEETS—SHEET 6.

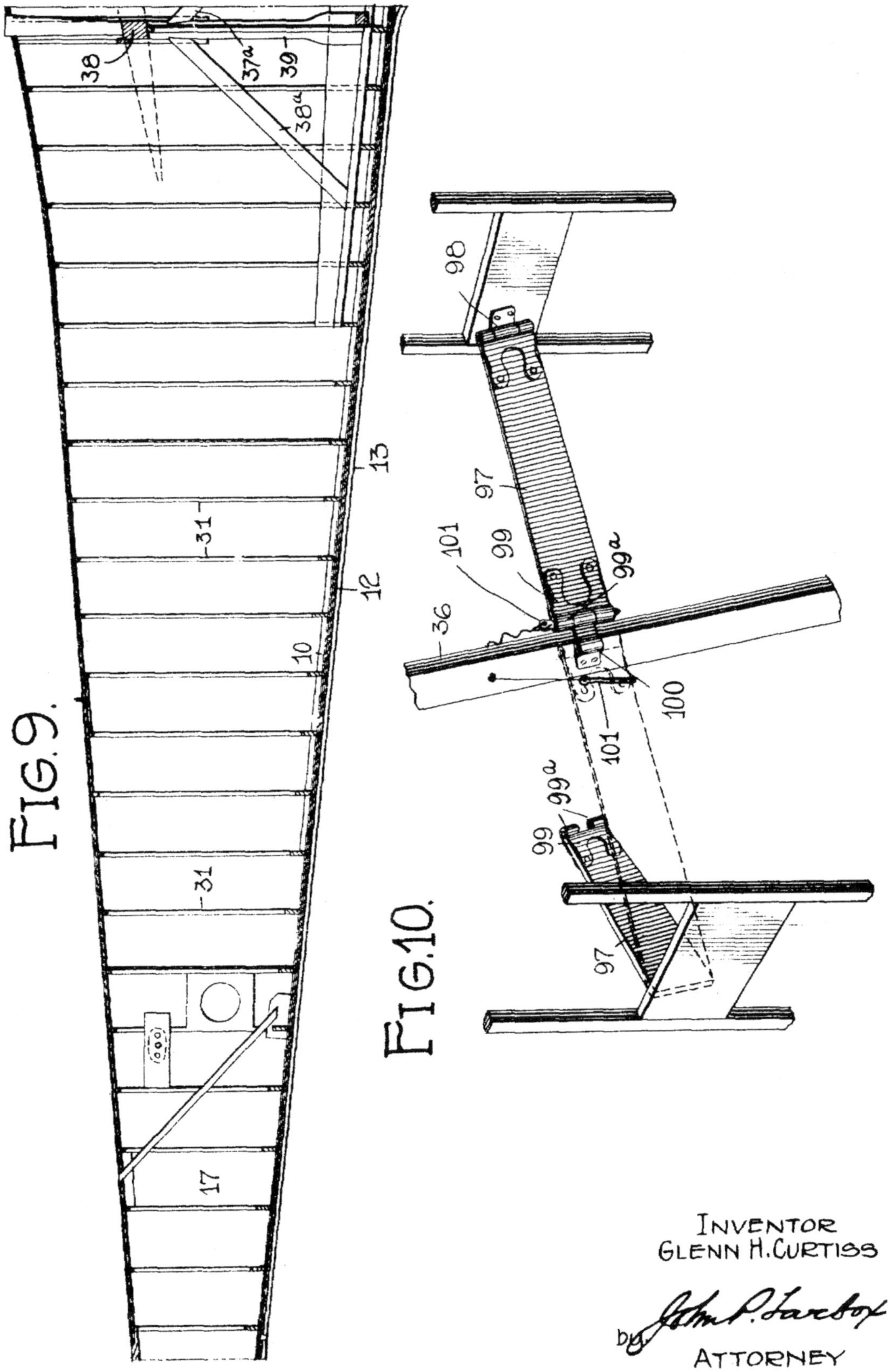

FIG. 9.

FIG. 10.

INVENTOR
GLENN H. CURTISS

by John P. Tarbox
ATTORNEY

G. H. CURTISS.
FLYING BOAT.
APPLICATION FILED DEC. 6, 1915.

1,246,016. Patented Nov. 6, 1917.
6 SHEETS—SHEET 3.

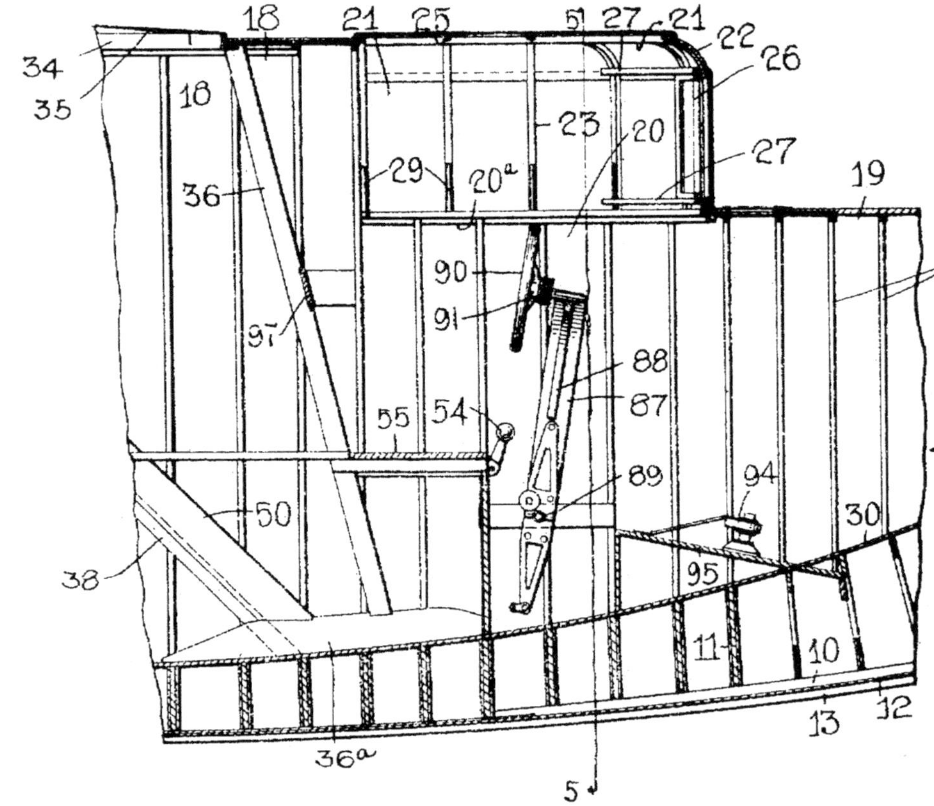

FIG.4.

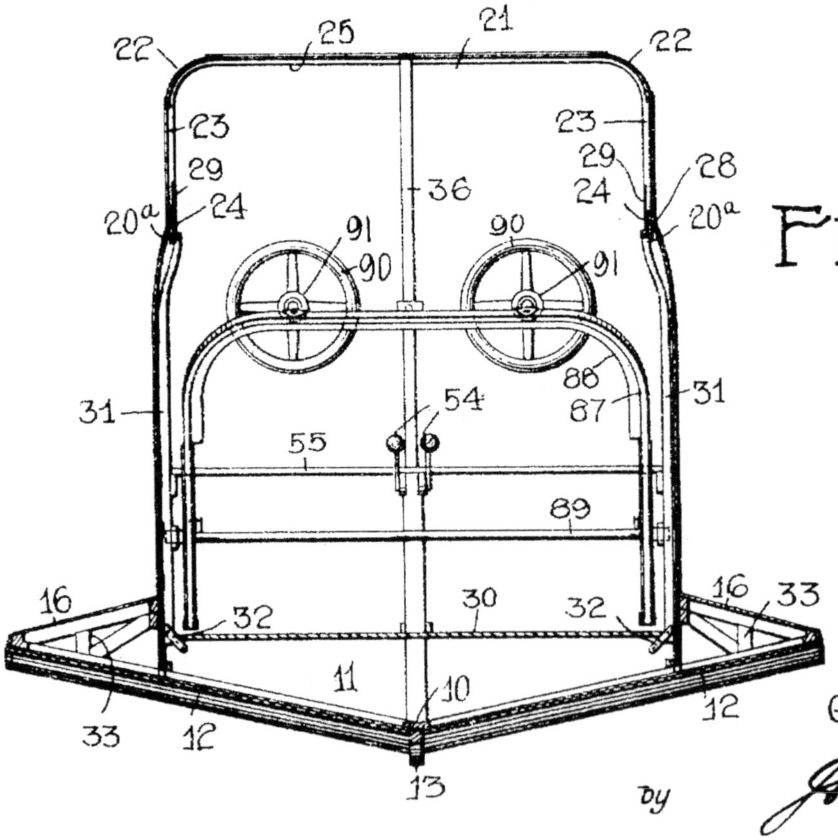

FIG.5.

INVENTOR
GLENN H. CURTISS
by
ATTORNEY

G. H. CURTISS.
FLYING BOAT.
APPLICATION FILED DEC. 6, 1915.

1,246,016.

Patented Nov. 6, 1917.
6 SHEETS—SHEET 4.

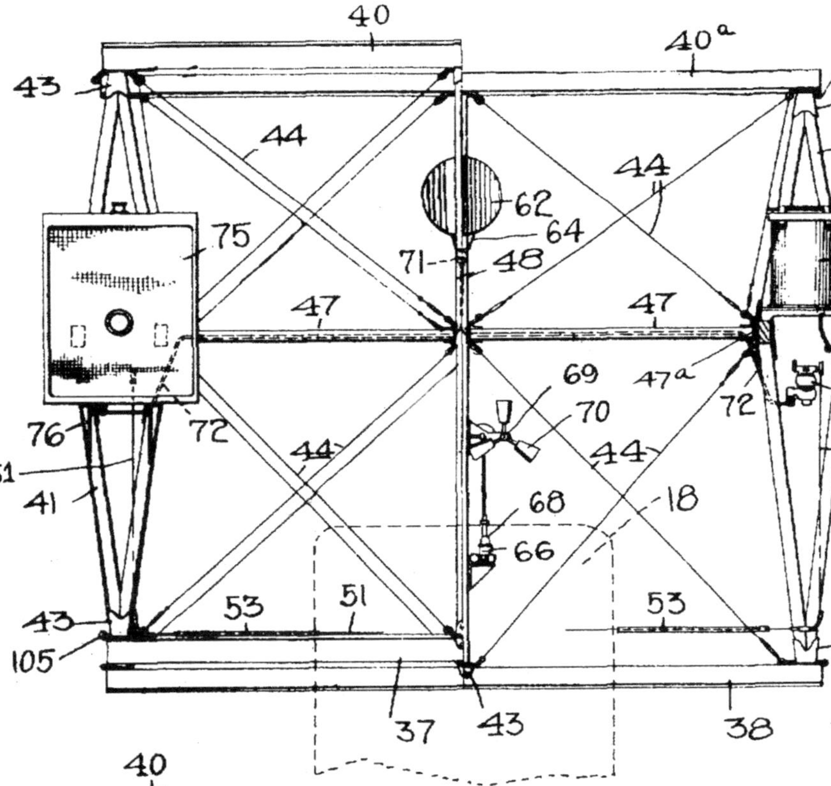

FIG.6.

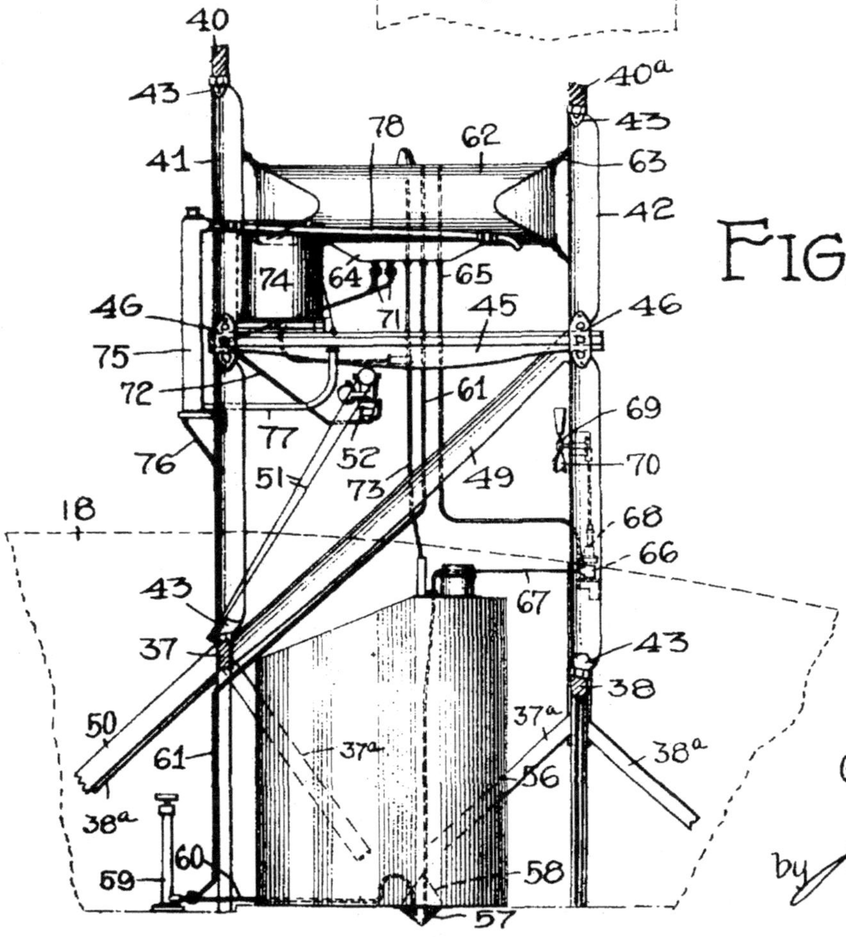

FIG.7.

INVENTOR
GLENN H. CURTISS

by [signature]
ATTORNEY

G. H. CURTISS.
FLYING BOAT.
APPLICATION FILED DEC. 6, 1915.

1,246,016. Patented Nov. 6, 1917
6 SHEETS—SHEET 5

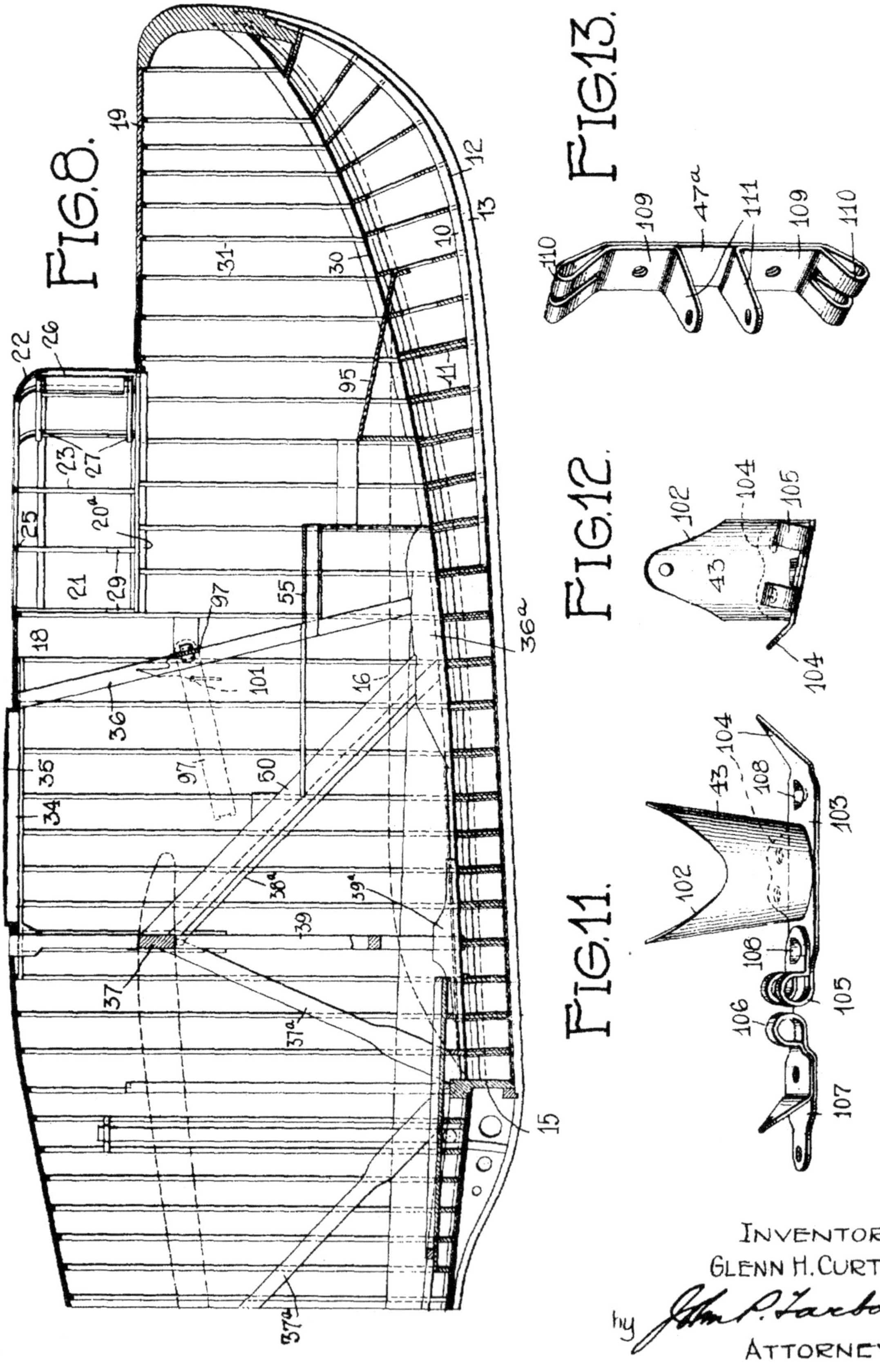

INVENTOR
GLENN H. CURTISS.
by John P. Tarbox
ATTORNEY

Right: "H-12 wings assembled." 18 October 1916. This early H-12 in the factory has been assembled without the engines being fitted to the airframe.

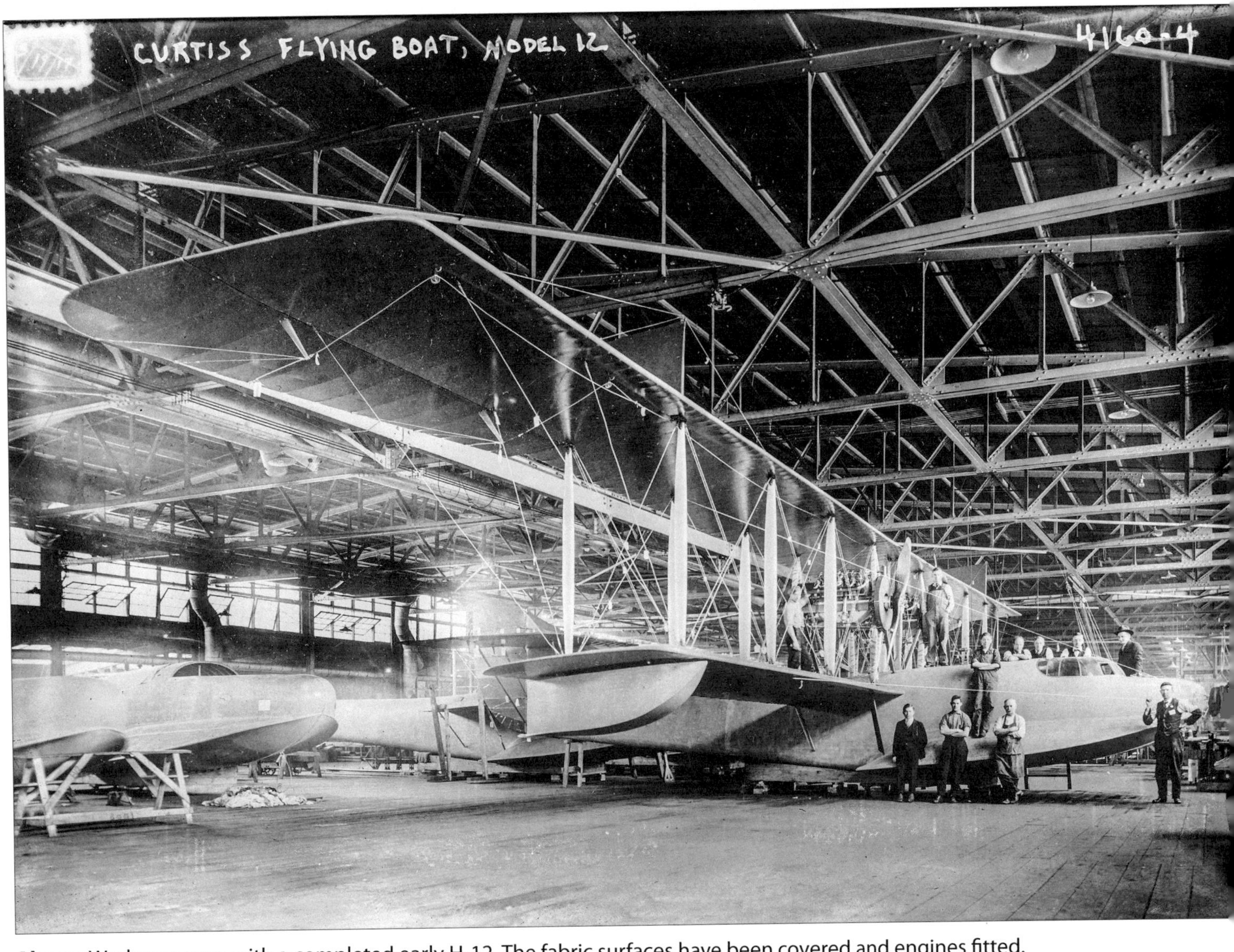

Above: Workmen pose with a completed early H-12. The fabric surfaces have been covered and engines fitted. 03.04.1917. Three unfinished hulls are visible in the photograph.

Above: Nose detail of a Curtiss H-12 under construction.

Left & Above: Detail of the engine mounts on an H-12B, 17 January 1918. Curtiss Aeroplane Co, Buffalo.

Above: Trial assembly of an H-12B at the Curtiss Aeroplane Co's Buffalo plant.

Below: Detail of the tailplane of the above aircraft. 11.09.1917.

Right: Detail of the tailplane of the preceding aircraft. 11.09.1917.

Above: Assembling an H-12.

Chapter 1 Endnotes

1. Wanamaker would continue to fund the progress of aviation, providing the backing for Adm Byrd's proposed trans-Atlantic flight from the US to Paris in 1927 amongst others.
2. Bass earned his captaincy in the Texas Rangers, a title he used for the rest of his life.
3. First Statement made 1st August, 1917, Commander John Cyril Porte, P.3. Copy in The Royal Commission of Awards to Inventors, Claim No.580. Claim of Mrs Mynne Porte. (Hereafter:- Statement of Commander John Cyril Porte of 1st August, 1917.) Copy supplied by late Ray Sturtivant. TNA WO158/993.
4. Other sources state Porte was so impressed that he convinced his employer, White and Thompson Ltd to purchase the boat and the rights to build Curtiss designs. The Company became the Norman Thompson Flight Co on 4 October 1915.
5. Statement of Commander John Cyril Porte of 1st August, 1917. P.10. *Op Cit.*
6. *Ibid.* P.5.
7. *Ibid.* P.10/11.
8. Goodal, M.H. *The Norman Thompson File*, Air Britain, UK, 1995, P.77
9. Casey, L.S. *Curtiss – The Hammondsport Era 1907–1915*, Crown Publishers, NY, 1981.
10. According to Glenn Curtiss' testimony post-war, Porte was never employed by the Curtiss Company but his financial arrangement was with Wanamaker. Curtiss Aeroplane & Engine Corp, New York; Mrs Porte and the Norman Thompson Flight Company. TNA T173/602. The Specification for the America 1915 included the following as Point 12: "If Lieut. Porte is unable to be at Hammondsport drawings and details are to be submitted or his approval." This would indicate that Porte's expertise was valued by Wanamaker and he had an input into the design of all the *America* boats including the proposed boats. Curtiss Documents in Museum of Flight, Seattle.
11. George E.A. Hallett was selected as the co-pilot after Towers departure. Casey, L.S. *Op. Cit.* P.191. F.N. Hillier recalled that Hallett was an officer in the Merchant Navy and he could "well remember his cool and balanced approach to this venture which, as we now know, must have been quite hopeless." "A Forgotten Pioneer." Letter to *Aeroplane*, 16 Dec 1955.
12. Read by Mr R. Moritz acting for Mrs Porte before the Royal Commission on Awards to Inventors. TNA T176/602 Pt.3.
13. Porte's notes to Maj W.F. Vernon's Report – "Flying Boat Development 1919". The title has been changed by Porte to read: "Flying Boat Development at Felixstowe Air Station 1914-1919", and Maj Vernon's name crossed out. Judging from the corrections to the text Porte called the original two Curtiss boats, 950 and 951, the *Americas*, crossing out the "Small" that Vernon had in the title. He considered that the *Small Americas* started with the order from Curtiss for the 50 boats (3545-3594). Vernon/Porte titled the AMC order as the "Saunders America" (1228-1235) which leads to the conclusion that Porte considered the four ordered from Curtiss (1236-1239) as *America* boats. TNA MUN 8/35. In a contract document between the Admiralty and the Curtiss Co of Hammondsport, the 50 flying boats are called the "H4 Transatlantic flying boats." TNA AIR1/3342/15/226/270.
14. Cutting from A (*Aeroplane*?) 16.12.1955. Copy in RAFM J.M. Bruce Collection, Box 22. For a detailed study of the America see Foxworth, T, "Curtiss Transatlantic Effort of 1914; or Wanamaker's Wonderbird," *World War I Aero Journal*, No.134, P.38. This article contains many photographs of the construction and modifications of the original America, See also *World War I Aero Journal*, No.111, P.82.
15. First Statement made 1st August, 1917, Commander John Cyril Porte, P.6. *Op Cit.*
16. Sueter had been involved in the development of submarines and torpedoes and was of an inventive mind and was a progressive in the 'Battleship navy". With only one short break he was the Director of the Air Department of the Admiralty until 1917. A difficult subordinate he fell afoul of his superiors and he was appointed to command the RNAS in the Adriatic. He was promoted to Rear Admiral and then retired. He later became a Member of Parliament and was knighted for his services to his country.
17. Porte is incorrect in his recollection as he quoted a price in his letter of 01.08.1914. Goodal, M.H. *Op Cit.* P.74.
18. L.J. Seely, manager of the Curtiss Aeroplane Co, was quoted as saying that the "airboat that was shipped on the Mauretania was the old America somewhat changed," which supports this supposition. Newspaper clipping, no source, no date, in *Curtiss Flying Boat – Publication & the Transaction*. TNA T20/22/15.
19. Goodal, M.H. *Op. Cit.* P.74.
20. M.F. Sueter, 24.08.1914. *Offer of 2 Curtiss Flying Boats.* TNA T20/22/15.
21. *Extract from Letter from our General Agent at New York, in Curtiss Flying Boat – Publication & the Transaction.* TNA T20122/15. The British were upset at the amount of publicity the transaction generated, the file bearing the notation that "Curtiss have made the most of the advertisement." Minute for Director of Transport, 18.10.14. *Ibid.*
22. The actual designations of these early boats are not known for certainty until the H-4.
23. Post-war Admiral Sueter was giving evidence in the cases of claims made before the Royal Commission on Awards to Inventors and when asked if Porte had "pushed the boats" on him, testified that
I naturally asked how the boat was getting on and full particulars because I had some flying boats in the Service and I wanted more...I bought these machines previously like the

Donnet-Leveque and the Sopwith boat for the service, and I would have bought these (Curtiss) *boats if the reports was favourable. I asked Commander Porte whether they were worth buying and he said he thought so and I cabled to America and asked the price and they quoted a price and then I got it cut down and I eventually bought the machines.*

Sueter made the point that Winston Churchill, as First Sea Lord, wanted to develop an American aircraft company as the RNAS would be short of aircraft as qualified staff of the British companies were called up for service, and the Curtiss Company was the one selected. Details are contained in *Curtiss Aeroplane & Engine Corp, New York; Mrs Porte and the Norman Thompson Flight Company*. TNA T173/602.

24. Statement of Commander John Cyril Porte of 1st August, 1917. *Op Cit.* P.7.
25. Maj W.F. Vernon, *RAF, Development of Flying Boats*. Report in TNA FILE MUN8/35. Contains Porte's amendments and corrections.
26. According to a list prepared in September 1915 the four H-4 boats cost $90,000, spare parts $4,054.20 and instruments £122.08.00; propellers 8 with couplings (four each RH & LH) $560, & propellers 8 drilled ready for couplings (four each RH & LH) $512; and installing 'OXX' motors in lieu of 'OX' motors in 1238 & 1239, $560. TNA AIR1/305/15/226/153 Pt.II.
27. Telegram Seeley to Curtiss Co, 13.11.1914. TNA ADM116/1577.
28. Letter from Air Ministry to Royal Commission on Awards to Inventors, 03.07.1924. *Thompson, NA – Flying Boats & Aeroplanes*. TNA T173/468.
29. W. Capehart to Bu C&R, 15.11.1916. NARA RG 72 Box 385 File O-ZF-2 Vol.1.
30. Capehart to Bureau C&R, 15.11.1916. *Ibid.*
31. Grandjean was in the US from November 1916 to April 1917. He had become a Naval Aviator on 10 May 1916. During 1918–1919 he made journeys to Germany, Norway and Britain. During WWII he was interned by the Germans. He died in 1948. Balsved, J.E. Danish Naval History, http://www.navalhistory.dk/English/TheOfficers/G/Grandjean_AEV.htm
32. From Capehart to Bureau C&R, November 24, 1916. NARA RG 72 Box 385, File O-ZF-2 Vol 1.
33. *Ibid.*
34. Casey, L.S., *Op. Cit.* states that two H-4 boats were ordered by Spain. P.186.
35. Memorandum: Present status of matters in aeronautics; to Secretary Navy from Office of Naval Aeronautics, 4.03.1916. Copy in Naval Aviation Historical Center, Washington Navy Yard, DC.
36. Sturtivant, R & Page, G. *Royal Navy Aircraft Serials and Units 1911–1919*, Air-Britain Publication, UK, 1992.
37. Reynolds, Clark G. *Admiral John H Towers – The Struggle for Naval Air Supremacy*, USN Institute Press, 1991. P.91.
38. Letter from F.N. Hillier. Aeroplane, 16 Dec 1955. It is considered that Hillier's recollection may have been slightly exaggerated as most other recollections agree that the boats were experiencing many problems that were gradually worked out as Porte improved the design.
39. Casey, L.S. *Op. Cit.* P.191. This additional H-1 may be that referred to as "a third sister ship" by Foxworth. (*World War I Aero* No.134, P.44.) To further confuse the matter of designations applied to the *America* boats, a list of Contracts with Curtiss prepared in September 1915, under the list of Spares for the "*America*" H4 type is listed one E/1 hull complete with fittings for every six machines. Copy in AIR1/305/15/226/153 Pt.II.
40. Sturtivant, R & Page, G. *Op. Cit.*
41. Notes from TNA AIR1/2590 in RAFM J.M. Bruce Collection, Box 3.
42. Statement of Commander John Cyril Porte of 1st August, 1917. *Op Cit.* P.8.
43. Roskill, Capt S.W. Documents *Relating to the Naval Air Service Vol.1*, 1908-1918. Naval Records Society, UK, 1969, P.197.
44. The Admiralty undertook to take machines, engines and spare parts to the total value of 1,050 machines divided as follows:

530 90-hp Machines	$3,653,500
360 160-hp Machines	$4,008,000
160 "America" Type machines	$3,353,500
1,050 machines	$11,013,000

There was no thought in the contract as to when and where the machines had to be completed and Curtiss could not produce that many in any case. This document also contains details of the contract for the Curtiss Canada twin engines land machine.
Admiralty to Curtiss Aeroplane Co, 06.12.15. TNA AIR1/305/15/226/153 Pt.II.

45. TNA AIR1/3342/15/226/270, Statement of Aeroplanes and Spare Parts Purchased by the British Admiralty with the Curtiss Aeroplane Company for the Period from March 26, 1915 to July 31, 1916, gives a breakdown of the Curtiss Co, Hammondsport, contracts with the Admiralty and how it was changed with the Admiralty agreeing to purchase other aircraft at an equivalent cost for aircraft, spares and engines cancelled – in the main the cancellation of the 100 Curtiss Canada biplanes and the 20 Model T Triplane flying boats.
46. Pix (Hallam, T.D.) *The Spider Web*, Arms and Armour Press, UK, 1979, P.21. Hereafter called Hallam, T.D.
47. Maj W.F. Vernon, RAF, *Op Cit.*
48. Secret Memorandum: Air Operations, P.3. May 1915. TNA AIR1/2538.
49. Page had been taught to fly by the Orville Wright at Dayton in 1912, and had worked as an exhibition flyer. Rejected for service by the Army on account of his age (he was born in November 1885), he joined the navy as a flight instructor in 1917. He was undergoing further instruction

at Felixstowe when he crashed. To understand the impact that his death had on the Felixstowe flying crews see Shirley, N. (Ed). "Wartime Memoirs of John Jay Schieffelin," *Over the Front Journal*, Vol.9 No.2, P.105. Schieffelin's view of his British contemporaries should be compared with those of Harold Wilcox, another USN airman who formed opposite views. See M.W. Knapp, "Flying Boats over the North Sea: 1918," *Over the Front Journal*, Vol.9 No.2, P.125.

50. First Statement made 1st August, 1917, Commander John Cyril Porte. *Op Cit.* P.9.
51. *Ibid.*
52. K. Molson describes the H-7 as having a short hull with the tail surfaces carried on booms and was in the configuration of the later NC boats. Molson, K. "The Felixstowe F5L," *Cross & Cockade Great Britain, Journal*, Vol.9 No.2, 1978. P.49.
53. This section is based on Casey's chapter on the H-7. An undated news report gives the first flight of the Curtiss giant "*Super-America*" or Model H-7, as 24 April at Newport News. "The machine is modelled on the first "*America*," that was developed two years ago," indicating that the year is 1916. The photograph accompanying the article is the same as that that Casey uses to illustrate the H-7.
54. According to E.R. Johnson, *American Flying Boats & Amphibians – An illustrated History* (McFarland & Co, USA, 2009), testing did not recommence until May 1916.
55. There appears to be no official record of an H-7 being accepted by the USCG. A news report in the *Aerial Age* for 1 May 1916, states that an H-7 "was held out of a consignment for the British and delivered to the Coast Guard with the personal compliments of the manufacturer, Glenn H Curtiss." As far as is known the British did not order any H-7 boats.

    "A heavy wireless equipment is to be installed with the idea that the boat shall be used as a scout to locate derelicts and either blow them up or stand by, even on a rough sea, while the wireless summons a cutter." The photograph accompanying the article shows the H-7 with the caption "The H-7 Model Curtiss Flying Boat which made a flight lasting half an hour at Newport News last week with eight people aboard." Copy in RAFM J.M. Bruce Collection Box 72. No H-7 flying boats are known to have been ordered by the British, and this report carries the fanciful journalism of the day when referring to the H-7 "blowing up" derelict ships.
56. "Curtiss Flying Boat Record." *Aerial Age Weekly*, 08.05.1916. The report states that the Allies are using 120 machines of "exactly the same type" but the US military had never even tried one out. Copy in RAFM J.M. Bruce Collection Box 72.
57. Was Capehart using the British designation of H-4 or was this a Curtiss designation and if so where does it fall in the history of the *America* boats? From Capehart to Bureau C&R, November 24, 1916. NARA RG72 Box 385 File O-ZF-2 Vol.1.
58. Hayward, C. *Practical Aviation*, American Technical Society, USA, 1919.
59. Reported in "Eddies", *Flight*, 10.12.1915, P.971.
60. J.L. Callan to G.H, Curtiss. Report on Condition of Trans-Oceanic Co. Cruiser. Callan papers, Bibliographical Section, Library of Congress.
61. "American Air Transport Services", *Flight*, 31.03.1921, P.230.
62. In his article "Sea Wings – Curtiss' Beautiful Biplane Boats – From America to PH-3 " Pt.1, (*Wings*, date unknown) Bowers states that the H-8 was allegedly built in 1916 according to Curtiss photo caption dates.
63. Curtiss Aeroplane Co to Bu C&R via Lt W. Capehart, 25.11.1916. NARA RG72 Box 385 File O-ZF-2 Vol.1.
64. Capehart to Bu C&R Subject: New Curtiss H-14 Flying Boat. 25.11.1916. *Ibid.*
65. First Endorsement to above letter. Signed Taylor. 02.12.1916. *Ibid.*
66. Telegram. From Curtiss Aeroplane Co to Bu C&R, 23.08.1917. NARA RG72 File O-ZS-17.
67. The Standard Model D Twin Hydros, serials 364-395, being cancelled on July 25, 1917. E-mail from Bob Casari 31.01.2005.
68. War Dept to Sec Navy, 24.08.1917. NARA RG 72 Box 385 File O-ZF-5.
69. Chance Vought has also been credited with the design of the H-8 for the British Admiralty. He came to work for Curtiss in 1915. The "Design of H 8 T.B. – Experimental, Report of Preliminary Lay-out bears the notation "From: Chance M. Vought." According to Casey this was for the tractor H-8. Copy in RAFM J.M. Bruce Collection Box 10. See also Schoeni, A. "Sixty Years of Vought," *Aeroplane Monthly*, UK, Vol.5, July 1977, P.367. This does not mean that Porte should not be known as the designer as all aircraft are the subject of a team working on different problems associated with the design such as stressing the wings, construction of the hull, etc.
70. Statement of Commander John Cyril Porte of 1st August, 1917. *Op Cit.* P.9. Here Porte is claiming that he was the designer of the machines that would become known as the H-12 *Large America*. There must have been those in authority who would have known of this design or else he would not have had the authority to order 50 of the new boats. When Porte made his statement the facts would have been easy to check.
71. Copy in RAFM J.M. Bruce Collection Box 9. This was included with other material sent by Louis Casey of the Smithsonian NASM to Prof E.A.G. Robinson.
72. Maj W.F. Vernon, RAF, *Op Cit.*
73. In their claim for compensation from the British Government after the war the Curtiss Company listed the boats supplied to the British as the Americas, H-2, H-4, H-8, H-9, H-12 and H-16.
74. Maj W.F. Vernon, RAF, *Op Cit.* Vernon considered this

"the largest single advance" made in the development of the flying boat.

75. Sturtivant & Page *Op. Cit.* state that all the boats on the order, with the exception of 8650, were delivered as H-12 boats, but this has not been confirmed. The mention of an H-9 delivered to the British would seem to indicate that these modified H-8 boats were known as the H-9 to Curtiss, but this is supposition.
76. Nicolson, D. "Design & Construction of Flying Boat Hulls." *Transactions of the Institute of Engineers and Shipbuilders of Scotland*, Vol.6, 29.04.1919, P.277. Photocopy in RAFM J.M. Bruce Collection Box 40.
77. The Report *His Majesty's Naval Aircraft built, building or under repair, March 1918* (Corrected to 27th February). (Admiralty, Air Department, March 1918), records 8650-8699 as H-12 boats, and N4330–N4350 as H-12B boats. Note that the last three of this order have been deleted at this date. There were 18 boats from each batch in service, that is 36 H-12 boats in commission at the 27 February 1918.
78. Maj W.F. Vernon, *Op. Cit.* Also The USN's Chart "Characteristics of Flying Boats and Seaplanes," of May 1918 only lists the H-12B with 330-hp Liberty.
79. Maj W.F. Vernon, RAF, *Op. Cit.*
80. Maj W.F. Vernon, RAF, *Op. Cit.*
81. Letter from J.M. Bruce to Colonel Gallop, 14.11.1959. Copy in RAFM J.M. Bruce Collection.
82. Comptroller General of Equipment to OC Plane Repair Park, Islington. 17.09.1918. Copy in J.M. Bruce Collection RAFM Box 7.
83. For example in September 1918 the proposed output of British flying boats was:

F.2A:	Airco	2
	Saunders	11
	Curtiss (H-16)	10
F.3:	D Kerr	4
	Phoenix	2
	Malta Yard	1
	Shorts	6

Notes by J.M. Bruce from TNA file AIR 1/500/15/331/4 Dept of Aircraft Production – Reports 79, 80 & 84. Sept–Oct 1918. RAF Museum J.M. Bruce Collection Box 7.
84. The memorandum noted that the *Large America* Type had 500 H.P., a Speed of 60 knots, a Radius of Action of 350 miles, an Armament of machine gun with either four 65-lb bombs or a 2 pounder Davis Gun. The number in commission was 5, with 4 being erected. 72 remained on the order to be delivered with an expected delivery of three per fortnight. TNA ADM1/8478/10.
85. This section on the Curtiss claim has been written from material in TNA T173/184. Curtiss Manufacturing Co – Aircraft – Hydro Aircraft. This file contains a sub-file dated 27.01.1923.
86. Wood & Another V The Aircraft Manufacturing Co and Mr Holt Thomas. The outcome of this action is unknown to the author.
87. In the Counter Statement submitted in the claim of Mrs Mynnie Port to the Royal Commission of Awards to Inventors, it was stated that "So far as can be ascertained the following types of flying boats have been constructed:

Small America	94
"Porte" Boats	11
H.12	74
H.16	75
F.2A	180
F.3	158
F.5	46

The Porte I and "Felixstowe Fury" were experimental boats of each of which only one was built."
88. There are no British records of an H-9, this being one of the missing Curtiss H series. It may be supposed that the modified H-8 machines provided to complete that order could have been redesignated as H-9 boats, but this is pure speculation at this stage.

Below: 8660's hull displays damage to the rear of the starboard fin. This is probably the damage that led to the machine being deleted on 14 November 1917. It was returned to service with an F.2A hull as an H-12 Convert.

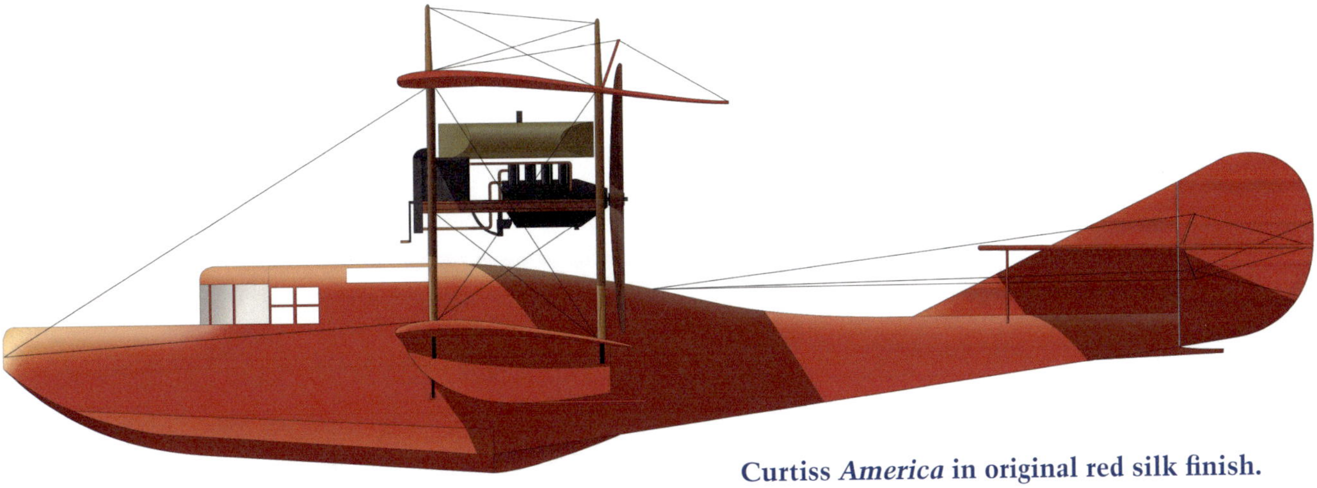

Curtiss *America* in original red silk finish.

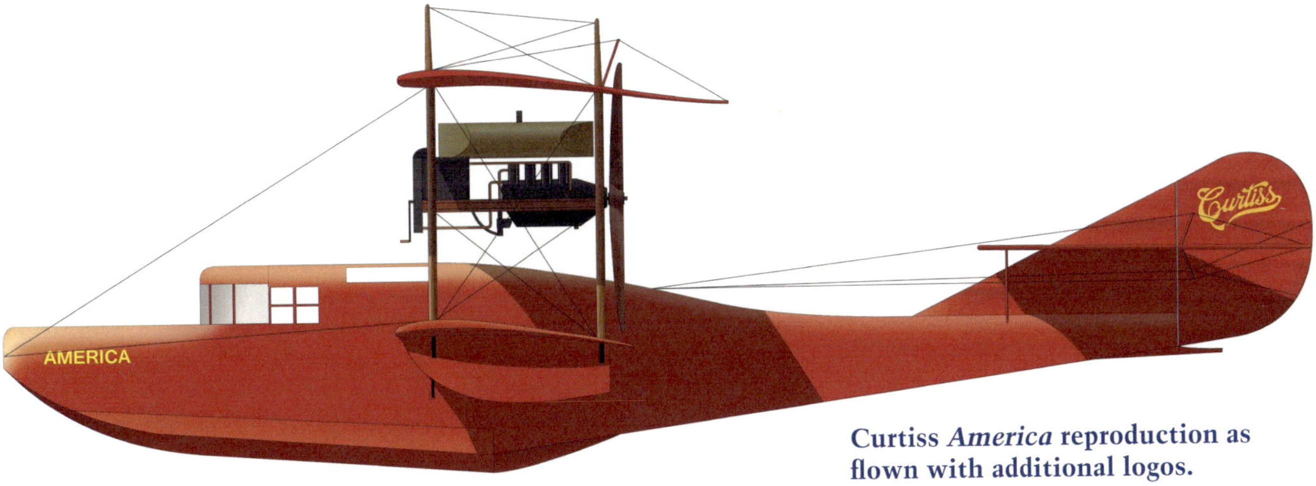

Curtiss *America* reproduction as flown with additional logos.

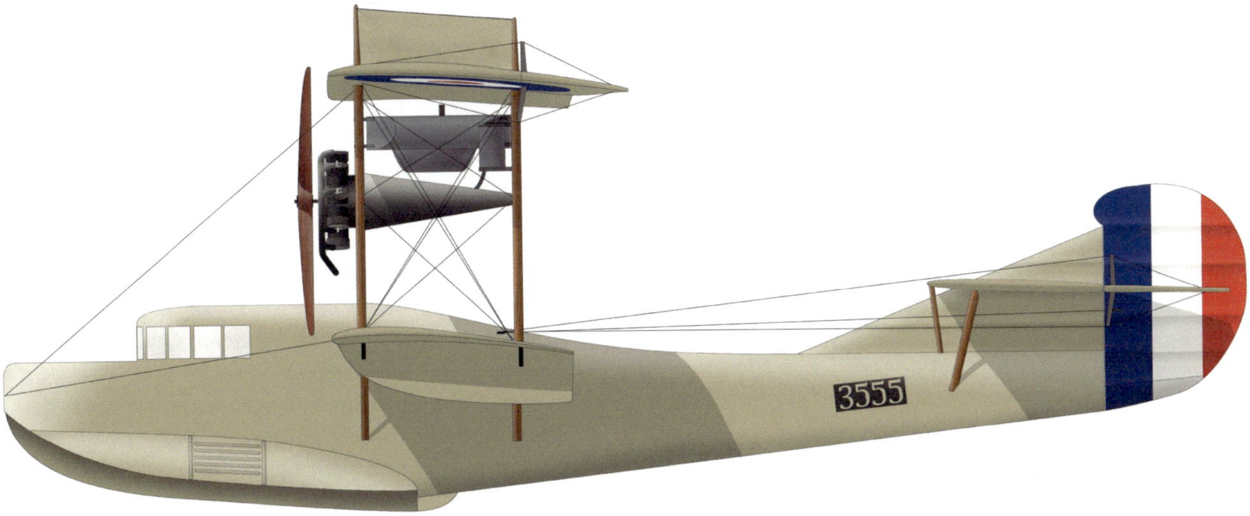

Curtiss H-4 3555 shown re-engined with Anzani radials. Delivered to Felixstowe in September 1915, it was at Calshot during 1915 and recorded in February 1916, and was deleted in February 1917.

2. Development of the Felixstowe Flying Boats

Above: FBA flying boat 3646 was built in France and delivered to White City in January 1916. The FBA was an example of the pre-war type of flying boat. The type was continually developed during the war years and saw service with the French, British, Italian and US Navies..

The Felixstowe flying boats were developed at the RNAS Seaplane Experimental Station, Felixstowe by John Cyril Porte from the Curtiss flying boats into the superb F.2A boats that carried the anti-submarine war into the North Sea. Despite their many shortcomings, the Felixstowe boats were practical machines that served successfully in the war and for a goodly time afterwards in the UK and the USA. There arose two schools of thought on the construction of flying boats in the UK, one that supported the Porte method of construction and the other that supported the ideas of Linton Hope. A series of lectures with question and answer sessions given to the Royal Aeronautical Society by exponents of both methods of construction have left a rich vein of historical material, with a deep undercurrent of a desire to discredit John Porte, that has formed the basis for this chapter.

As noted in Chapter 1, Porte had been invited to the USA to be the pilot of the Curtiss-Wannamaker *America* flying boat that was to fly the Atlantic in response to the *Daily Mail's* offer of a prize for the first crossing by an aircraft. With the outbreak of war, the flight was postponed and Porte returned to England where he volunteered for the RNAS. He was appointed Squadron Commander at Hendon.

> *Directly I got back I saw Commander Suter* (sic) *the head of the Air Department.*
>
> *I saw him the next morning after I arrived and I also informed him that there were these two flying boats which we had built in America to fly the Atlantic which were available and could be bought by the Admiralty if they desired to buy them....*

The boats were delivered to Felixstowe and Porte, who was at the time in charge of Hendon, was to test the first one. Commander Seuter came down and after a flight with Porte reported to the First Sea Lord, Winston Churchill, that he considered them excellent and Churchill then ordered that 12 be contracted for at once.[1]

In order to understand the experiments and development of the Felixstowe F boats, it is first necessary to understand the difference between a landplane's and a flying boat's requirements. The following has been taken from the lecture given by Major J.D. Rennie, who had been the Chief Technical Officer while Porte was in command of RNAS Station Felixstowe, to the Royal Aeronautical Society and published in their Journal in 1923.

The usual biplane configuration of a flying boat differs from that of a large aeroplane in that the span of the upper wing is considerably longer than the lower. This is not intended primarily to obtain greater aerodynamic efficiency owing to the absence of biplane interference on the extensions, but to reduce the wing area in proximity to the surface of the water, thus minimising the risk of damage to the lower wing in a rough sea. For the same reason ailerons are not fitted to the lower plane.[2] To allow for adequate lateral control the ailerons should be positioned as far outboard as possible.

The position of the wing tip floats should be as close to the

Above: The Admiralty Air Department A.D. flying boat used the Linton Hope method of construction and represented the future type of wooden construction that was to be used on the first Supermarine Southampton flying boats post-war. Supermarine received a contract for A.D. boats, the hulls being delivered to this firm where the wings, etc., were added. The prototype 1412 is illustrated here. Supermarine sold versions of the A.D. Boat as their Channel seaplane post-war. (For the story of the A.D. Boats see Owers, C.A. *AD Flying Boats*, Albatros Productions Datafile No.159, UK, 2013.)

hull as possible so that less shock is transmitted to the wing structure on takeoff, landing and rolling, one wing down. The result of these considerations is the wing arrangement described above.

"It is said of the "F" boats that the use of stabilisers (anti-skid fins) on the top plane is a very inefficient aid to lateral stability. They were never really intended as such," the top plane extensions requiring bracing to withstand downloads that are applied at high speed flight or inertial load during a bad landing. The rectangular cross-braced kingposts were considered to be better than the triangular type, and by fairing it in, it also acted as a stabiliser.

"In the F boats, which had twin tractor airscrews, about three-quarters of the tailplane was in the slipstream and seemed fairly satisfactory, but on trial the boat was decidedly tail heavy, as were all of the F boats. This was partly due to the fact that when carrying full military load the CG was further back than was originally intended. With pusher propellers the conditions are simpler but they are very liable for damage from parts that may work loose, or even tools that have been left through carelessness.

"With regard to fin and rudders, these are of relatively large area due to the long forebody of the flying boat. The rudder area of the F boats was barely sufficient for control with one engine out.

"As with the aeroplane, controllability at low speeds is of importance, but probably less so as generally the extent of a landing ground is not so restricted. Also once alighted the boat pulls up quickly due to the large hull resistance. Thus it is possible to glide at a comparatively high speed until close to the water before flattening out and alighting."[3]

The modifications carried out on and the new type hulls evolved at Felixstowe were arrived at from full size experimentation as model hull testing was not available at the time with the exception of the Fury triplane. From 1915 experiments with full sized machines were carried out to correlate the results of tank tests at the National Physical Laboratory's William Froude tank. These were not successful due to the problems of collecting data. Visits were made to seaplane stations for the purpose of studying the behaviour of the machines in disturbed water when getting off and alighting. The following visits took place:

> July 1916 at Felixstowe when the "America" flying boat and Short seaplanes were flown. The machine was described as the 4,000 lb America type with propellers rotating in the same direction.
>
> *Reputed getting of speed 38 knots, speed in flight 48 knots. The*

machine rose and settled three times. The pilot held the tail down on the water and bow up until 30 knots was reached then decreased the angle so that the tail came off, after which the flying speed was quickly reached. To lift the tail earlier than 30 knots would mean throwing up a lot of water at the fore end and it could not be lifted earlier than 18 knots. The machine took a little time to reach the speed of 30 knots, but when the tail was lifted accelerated rapidly. This agrees with tank results which show that the drag of the tail adds very largely to the resistance. The landing was gentle on each occasion, once on the tail tip and twice on the step. The machine was run up once, for a few seconds, at a speed slightly over 30 knots tail out, without control. The air balance was fairly good and the machine moved but very slowly from its longitudinal trim.

September 1916 at Calshot where the F.B.A. flying boat was flown.

August 1916 to Southampton to view the A.D. flying boat.

October 1916 at Felixstowe saw some data obtained on a Porte boat.

A "Porte" boat with chine running continuous from stem to stern, front step a little abaft the centre of gravity and two other steps at the rear and under the chine. On practically smooth water and with no wind, both with the elevator full up and down, failed to unstick the tail. With no observer, less petrol, and a little wind, this machine got off, but it was too far away for its behaviour to be noted.

"The great drawback in all these cases was the absence of any reliable scientific data and the contradictory character of the *opinions* expressed by those concerned in the running of the machines. It was in no case possible to get reliable speeds or inclinations of the machine. The visits served, however, to show the general character of "porpoising" and to some extent the means adopted by pilots in evading this phenomenon."[4]

It was hoped that by using models in the Tank it would be possible to save a certain amount of experimental work at the stations with greater ease and considerable savings. Until the results of tank testing and full sized boats could be established the only way to determine the best type of hull was by full scale experimentation and this was how Porte approached the problem.

At rest the total weight of a flying boat is supported by hull buoyancy, and lateral stability by means of the wing tip floats. Owing to the high centre of thrust and low water and air resistance up to speeds of say 10 knots, the throttle is opened slowly, and elevators held up to prevent trimming by the bow. As speed increases, elevators are put into neutral when the hull should trim back of its own accord. From this speed the load that is supported by the buoyancy is gradually transferred to the hydrodynamic water forces acting on the planing surface, the fore body rising first, followed by the tail, which may be assisted by putting the elevators down slightly until the boat is planning cleanly. Water resistance will increase steadily, however if the hull is designed correctly, it will continue to drop owing to the improved working to the step and to the increasing percentage of the load taken by the wings. The speed at which the water resistance reaches a maximum is generally known as the "hump" speed. Generally a boat that gets over the hump has power for flight. From this speed or generally a little above, a sharp pull up on the elevators will make a clean break away from the water, the boat becoming entirely airborne. The elevators are then depressed before trimming to gain height.

The hull should be designed so that it is able to be trimmed back to such an attitude that the angle of incidence of the wings is that corresponding to the lowest safe speed, usually a few knots above stalling speed, without excessive elevator movements and increase in hull resistance. Serious accidents due to taking off in a stalled condition are rare as flying boats are generally well above the lowest flying speed providing sea conditions are suitable.

Up to about the "hump" speed lateral stability is obtained by use of the wing tip floats and such aileron control as is available. After the "hump" speed the boat trims naturally onto an even keel due to the stable hydrodynamic forced acting on the planning surface. As speed increases the aileron control becomes more effective.

Thus a flying boat hull must

(a) Avoid diving at low speeds.

(b) Have seaworthiness.

(c) Have hydroplaning efficiency and landing with minimum shock.

(d) Have stability at high speeds on the water.

(e) Have the ability to trim fore and aft to enable take offs and alightings.

As many readers may not be as familiar with the terminology applied to the construction of flying boat hulls, the following definitions are used (see table on facing page):[5]

As noted in Chapter 1 the Admiralty bought the two Curtiss America flying boats that had been built for the proposed pre-war trans-Atlantic flight. These boats were delivered to Felixstowe in October 1914 where they underwent trials. The boats proved promising and, as previously recorded, Churchill, First Sea Lord, ordered a dozen as the Curtiss H.4 with more powerful engines.[6]

Porte has recorded that he wanted to build the boats in the UK and so an order for eight (1228–1235) was given to the Aircraft Manufacturing Company. These were designated as H-4 *America* flying boats. Four were ordered at the same time from the Curtiss Aeroplane & Motor Co Inc (1236–1239).[7] As noted further on it appears that Curtiss provided the AMC with details of the design. The four Curtiss boats arrived long before the ones to be made in England were completed. These latter boats were built under Porte's supervision at Hendon. The hulls were built by Saunders at Cowes. Later orders followed with 50 H-4 boats being supplied by Curtiss under Contract No. C.P.01533/15.[8]

Porte had flown the H.4 and knew of their limitations and began the task of modifying the boats to get one that would be usable in the North Sea. "At the end of January 1915 I was

Flying Boat Hull Terminology	
Athwart	Transversely, at right angles to fore and aft.
Beam	The transverse member to support the deck, and the sides. The breadth of a boat.
Bottom	Usually understood as the part of a boat below the waterline.
Chine	Where the top sides meet the bottom at an angle.
Feather edge	When a plank or timber tapers to a very thin edge, "Tapering to nothing."
Floors	Transverse members connected the heels of frames to bottom and keel.
Forebody	The fore part of a boat which is forward of the midship section.
Frames	The transverse members to which the skin is fastened.
Ribs	The frames or timbers of a boat.
Scarf	A method of joining pieces of wood or metal by tapering their ends. A box scarf is when the ends are not tapered, but a half thickness cut out of each part so that when put together the parts form only one thickness.
Seaworthiness	Seakeeping qualities, with special regard to the ability to takeoff and alight in smooth or moderately rough seas, keeping the cockpits dry, airscrews clear of solid water. Also, in addition, such matters as reserve buoyancy, wallowing in a following sea, and easy riding at moorings.
Stringer	Strengthening fore and afters connected to frames or timbers.
Sternpost	The vertical member where planking terminates aft.
Timbers	The transverse ribs of a boat.
Hull	The boat as distinct from her superstructure.
Keel	The fore and aft members in a boat to which the frames and garboard strakes are fastened, or it a diagonal planked boat, where the planks end.
Keelson	An inside keel fitted over the throats of the floors.
Lines	A general term applied to the drawing or design of a boat as depicted by fore and aft lines and cross sections. A boat is said to have "fine lines" when she has a low block coefficient.
Moulds	The skeleton templates to cut the frames by, or to hold the boat in shape while the timbers and stringers are being fitted in place.
Planking	The outside skin of a boat; planks laid on the frames or beams of a boat whether inside or outside.
Transverse	Athwartships. At right angles to the line of keel.
Tie	Diagonal connecting and strengthening member.

appointed to take charge of Felixstowe Air Station and in May of that Year I left Hendon altogether and took over Felixstowe completely and have remained there practically continuously" since then. When he went to Felixstowe there were then only "50 men and three small sheds; there are now 1,000 men and an enormous number of sheds...I have built the whole thing up"[9]

Felixstowe's Experiments

When the first of the Curtiss flying boats arrived in the UK, "Commander Porte was specifically instructed by the Director of the Air Department Admiralty to experiment with and improve the said boats."[10]

"When the first experiments were taken in hand no engines of high power were available. Consequently, seaplane designers were faced with the problem of "Getting off" as the chief difficulty. For this reason the early experiments were made with the main object of developing hydroplaning efficiency and the question of simplicity of structure and shockless landing were neglected."[11]

The first hull tested was a modified Curtiss *America* flying boat serial 950, one of the original America boats purchased by the Admiralty. This hull weighed light 3,100 lbs and, on certain occasions, 4,500 lbs was taken off the water. The original hull was 30 feet long, and was modified by the addition of a wide longitudinal projecting fin forward, ending at the single step that was under the CG of the machine. The fore and aft angle between the underside of the tail and planning surface of the ship was 10°. The available engines gave about 160-hp.

At high speeds all single step hulls balance on the step and trim depends on the angle of the tail portion, which in order to avoid water drag, should be kept up while accelerating the boat. The original tail plane was lifting and hence excellent from this point of view, and as much load as could be flown with could be taken off the water. In order to improve stability in flight the tailplane was made negative. In smooth water there did not

Above: The "Incidence Boat", H-4 3546 with experimental hull. It is fitted with rotary engines in this photograph. Initially fitted with two 100-hp Clerget engines, it was re-engined with 100-hp Gnome Monosoupape engines. It was finally given two Anzani radials. It served from July 1915 to April 1917. Its part in the Felixstowe story is still unknown.

Above & Facing Page, Top: 3569 was fitted with a steep V-bottom with fins that carried on almost to the tail without a step. It proved incapable of getting off the water and a step was added as shown on Plan No.5. This enabled the machine to takeoff and exceptionally smooth landings were accomplished due to the deep V-bottom. It was recorded as being tested by Porte and Flt Lt R.J.J. Hope-Vere on 30 May 1916. It lasted until being deleted on 25 April 1917.

Above & Following Page, Top: Curtiss H-4 3570 still has its Curtiss engines installed. This experimental hull had a relative deep afterbody and reduced fin area. Compare with that of 3579. Note the curved shape to the cockpit that has windows in the roof. Compare with 3570. These photographs were included in the section on 3545 in Sueter's report. Unfortunately no photographs of 3545 have been located to date.

Above: Curtiss H-4 3579 with Porte type hull. Compare the flat appearance of the cockpit with that of 3570 at top. Note the torpedoes lying in the background.

Above & Below: Aircraft Manufacturing Co (AMC) version of the H-4 1231 was fitted with an experimental hull built by Saunders to Porte's design. The rounded aft section would be similar to that of 1230. Suction prevented the boat's ability to take off and was blamed on the shape of the hull.

Above: Porte Baby prototype 9800 outside the camouflaged hangars at Felixstowe with another boat with an experimental hull in the background. Note the small cockade on the underside of the Baby's upper wing.

seem to be any appreciable loss in hydroplaning efficiency, but in rough weather, owing to the lack of buoyancy forward, this hull was very wet. The nose of the hull tended to dig in and water was thrown up over the engines to such an extent as to cause the engines to misfire, thus indirectly making it difficult to get off.

A new hull was built for the Aircraft Manufacturing Company by S.E. Saunders Ltd in which the fins were narrower and carried further aft as the step was under the rear spar. The straight lines of the previous form were replaced by curves, the planning bottom being slightly hollow. The underside of the tail was rounded and construction was lighter. This machine was given the serial 1230. This hull proved inferior to that of 950, largely due to the rounded tail section that was not only a less efficient hydroplaning surface but increased suction making it very difficult to lift out of the water in calm weather. This hull was 32 ft 7½ in in length. While there was a savings of 300 lbs in weight over previous hulls due to its different construction, it was considerably weaker and "did not outlast many landings." The Saunders hulls were covered in the patented Consuta stranded copper wire sewn plywood and heavily varnished.

The next hull was made for Curtiss H.4 serial 3545 and was similar to 950 but the tail portion was 2 feet longer, and the fins were wider. The fore and aft angle was reduced from 10° to 7°. "Consequently the machine had to be held at a finer angle of incidence when planning to avoid drag due to (the) submersion of the tail, which caused the getting off speed to be higher." In smooth water the hull gave a better planning efficiency than the original hull. While the total weight of 4,500 lbs was taken off the water by this hull, the same load as for the original hull, this hull was 300 lbs heavier than the original one.

"The main conclusion arrived at from these experiments were that, from the point of view of planning efficiency and low getting off speed, a large fore and aft angle was essential and a flat bottomed tail portion was advantageous. But it was also evident that much remained to be done in the direction of lighter construction and of general seaworthiness." From the experiments with these hulls it was learnt that from the point of view of the ability to hydroplane and low take off speed, the tail portion should be flat bottomed to reduce suction, and the fore and aft angle should be large to obtain the necessary trim.

Next it was decided to tackle the problem of easy landing conditions, and increased strength of the hull without sacrificing planning efficiency. "This series ended with the production of 'Porte I', which machine showed marked improvement in these particulars."

In the early days all landing breakages of flying boat hulls occurred at the step, a structural weak point, indeed a *Small America* had broken in half at the step. The need to reconsider the best form of hull was evident. The next experiment was to find if a step was necessary at all. A complete new steeper V bottom was built on 3545 having no step but with a steeper fore and aft angle of nearly 20° and with the tail very much swept up with the keel line following a smooth curve. The fins that extended for ¾ of the length were extended back to meet the hull instead of ending off square under the wings as before. The result was to decrease head resistance by making the whole hull more streamline and also to make a stronger hull. This machine was given the serial 3569 and was known as the Transitional boat, as it was the link between the *America* boats, all those boats that had their fins cut off short at the step, and the Felixstowe group with their long fins and steep V bottom. This hull was 32 ft 2 in long. With the available engines (presumably of nominal 200-hp) there was not enough power to take off. A step was added 5 feet behind the CG and the hull was then capable of getting off but with only 4,200 lbs total load as against 4,500 lbs with the earlier hulls. Although "Getting Off" was not as good as the *America* boats, landing was exceptionally easy owing to the large fore and aft angle. The deeper V resulted in little or no landing shock with a normal or nearly stalled landing. "This latter method of landing is a very severe test because, immediately the tail portion touches the water, the heavy water drag pulls the machine down very suddenly." Owing to the step being so far aft of the CG, the hull ran at a small angle and required a large moment to trim the machine back for takeoff. To relieve the pilot of this load the step was shifted 3 feet forward.

The next experiment was the building of "Porte I" an entirely new single step hull built at Felixstowe to carry the same Curtiss superstructure as before. The bows were kept fuller and given a flare or concave entry. Porte used an entirely different form of construction for the next hull with longerons and spacers with wire bracing as used in the construction of landplanes, rather than the boat builder methods used for the Curtiss hulls. The *America* hulls depended on their tubular shape and stiffness of their skins for strength. They had reinforcement from the keel and, in the larger boats, from bulkheads, and some centre line wiring in the tail, but the rounded form and continuous planking was essential to the structure. Porte's method gave the requisite strength for a low weight. Originally a single step, located below the rear spar, was fitted. The hull was 36 feet overall in length, three feet longer in the nose and two feet in the tail than the third type of hull.[12] The fore and aft angle was 18° and the tailplane was raised 7 inches more than that of the fourth hull, and the line of the keel was kept higher. Fins were carried well at of the step and swept back into the hull. The V-bottom was similar to 3569 and the bows were fuller with a distinct flair on the first three feet.

With this hull the tail portion was liable to catch the water and drag held the speed down to below take off speed. To overcome this, a second step was added 7½ feet from the stern. This modification meant that it was now possible to takeoff but with less load than the earlier hulls. A third step was added intermediate between the main and aft steps. This allowed the load to be brought back to very nearly that of the earlier hulls. This hull was far superior to any hull previously tested at Felixstowe. Owing to the improved form of bow, the cockpits were perfectly dry. Landing shocks were reduced to a minimum, and behaviour generally in alighting and getting off was excellent. Seaworthiness was far superior to the preceding hulls. This hull was named Porte I and the aircraft was subsequently designated the F.I.

3580 had a long and varied career and at the Felixstowe Seaplane School by December 1917, and was flown for many hours by students who did their worse to her. When dismantled for overhaul no serious defects were found other than corrosion of various metal fittings. It survived until January 1919.

In 1918 she was re-engined with two 150-hp Hispano-Suiza engines and subject to further testing. With no step she just staggered into the air in a stalled condition showing that such a thing was possible with a low enough loading. With a single step she was not as efficient as with two steps, thus seeming to confirm the results of the Transitional Boat.

The decision was made to experiment with larger hulls and the Curtiss H-8 *Large America* serial 8650 was obtained. The H-8 was a larger boat built by Curtiss and one that Porte claimed that he had designed.[13] Porte was in Canada in 1915 to witness the testing of the Curtiss Canada twin engined landplane based on the Curtiss *America* flying boats.[14] Given his connections with Curtiss it is assumed that he would have taken the opportunity to visit the Curtiss manufacturing facilities and discuss his ideas for flying boats for operations in the North Sea. He so claimed in his post-war application for payment due to his inventions with respect to flying boats.

As detailed in Chapter 1 the H-8 was modified to become

Above: Rear view of the Porte I showing the V-bottom and how the tail of the fuselage like rear hull rose quickly above the water line.

Above & Below: 3580 had a new hull built to Porte's method of construction. Initially it has two Anzani radial engines but is shown with two 150-hp Hispano-Suiza engines. This machine became the Porte 1, later renamed Felixstowe F.1. The F.1 served at the Felixstowe Seaplane School from at least December 1917 to January 1919.

the H-12. With more powerful engines the H-12 boats did good work but their hull was structurally weak.[15]

Particulars of the H-12 were as follows: Hull length: 40 feet; Maximum beam: 10 feet[16]; Fore and aft angle: 7½°. The hull weighed 2,200 lbs and the complete machine 6,200 lbs when light. The intended takeoff weight was 8,700 lbs with the two 160-hp Curtiss engines. Unfortunately these engines proved incapable of taking this load off and 240-hp Rolls Royce engines were substituted. Even with these engines this load could only be taken off with difficulty. "This was due to lack of buoyancy and it was only when lightly loaded that "getting off" could be accomplished with ease."

There was a distinct "hump" speed of about 18 knots, due in part to propeller inefficiency at that speed but mainly to the fact that while at that speed the wings were lifting very slightly while the lift of the hull was also poor and its water resistance high. Once over this "hump" speed, there was little trouble in getting off. "It was evident that a greater load could be taken off were not for this phase of inefficiency."

The Porte I hull was superior to that of 8650, so it was decided to construct a new hull on the same lines to take the same *Large America* super structure of the Porte I. This was known as the Porte II (later F.2) and was designed along the lines of the Porte I as much as possible on a larger scale. The

Above: The H-8 was unsuitable for RNAS requirements and Porte soon produced a new experimental hull to 8650's aerostructure that led to it becoming the Porte II (later F.2). Note the enlarged fin.

Below: F.2 with top decking extended along to the stern post and standard fin.

Above: Saunder's built hull on 1231. The actual input that Porte had into the design of the AMC batch's hulls is unknown, however it must have been considerable given the different hull shapes tried out. They were built under his supervision.

Above: Porte hull on 3569. Note what appears to be a camera window on the port side bottom of the hull.

Above: Early H-12 hull with pointed bow and early canopy.

Above: The "Incidence Boat" 3546. What input, if any, Porte had into this hull is unknown.

Above: Probably 3579 with a standard hull showing the Curtiss shallow V-form with a slight concave planning bottom.

Above: The Curtiss H-8 hull still retained the concave bottom but was of greater V-form.

Above: The hull of 8651, the first Curtiss H-12 to arrive at Felixstowe. The hull bottom is concave and the canopy is different to later H-12 boats. Note the "Incidence Boat" in the background.

Above: The Porte I hull. When combined with the flight surfaces of 3580 and with two 150-hp Hispano Suiza engines providing power, became the Felixstowe F.1. Note the straight V-form of the hull bottom.

Below: The Porte II (later F.2) hull.

Experimental Flying Boat No.950

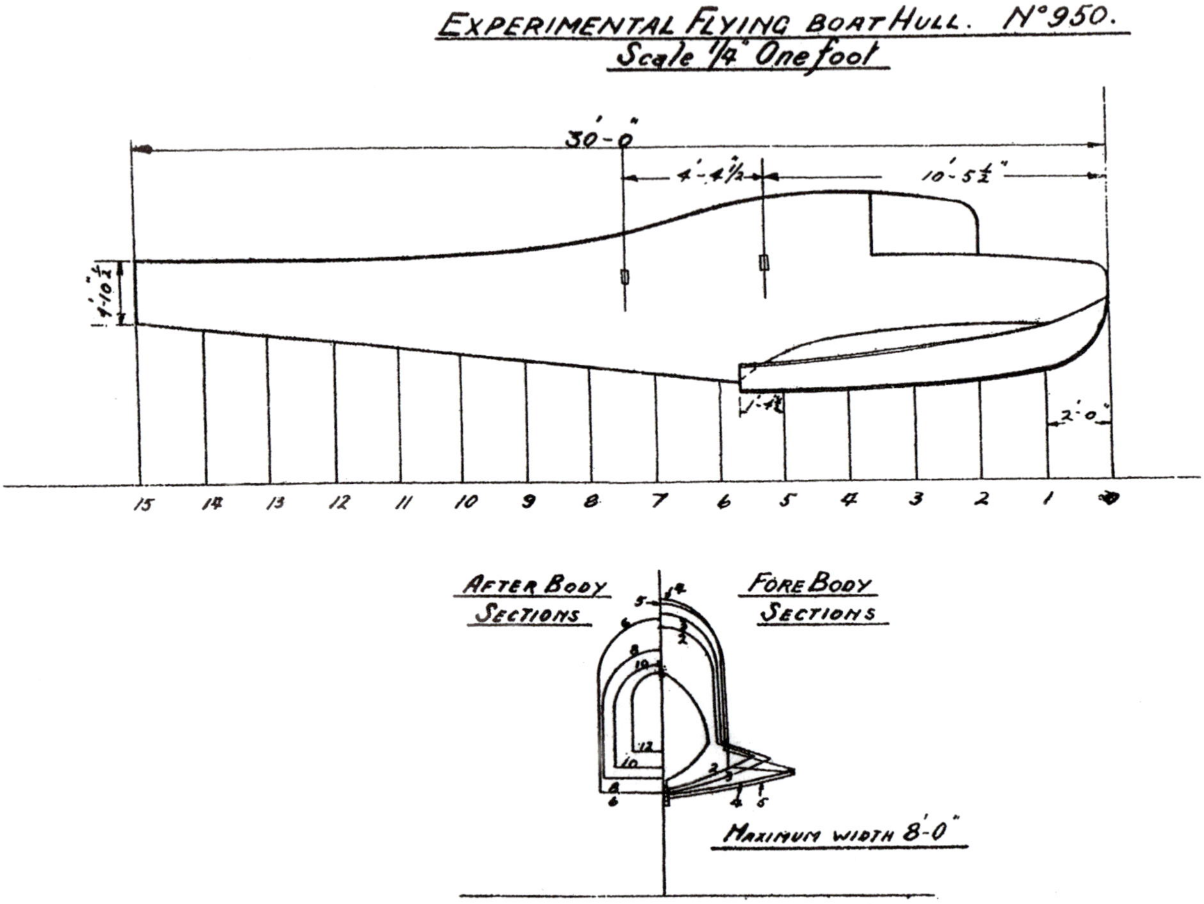

Experimental Flying Boat No.1230

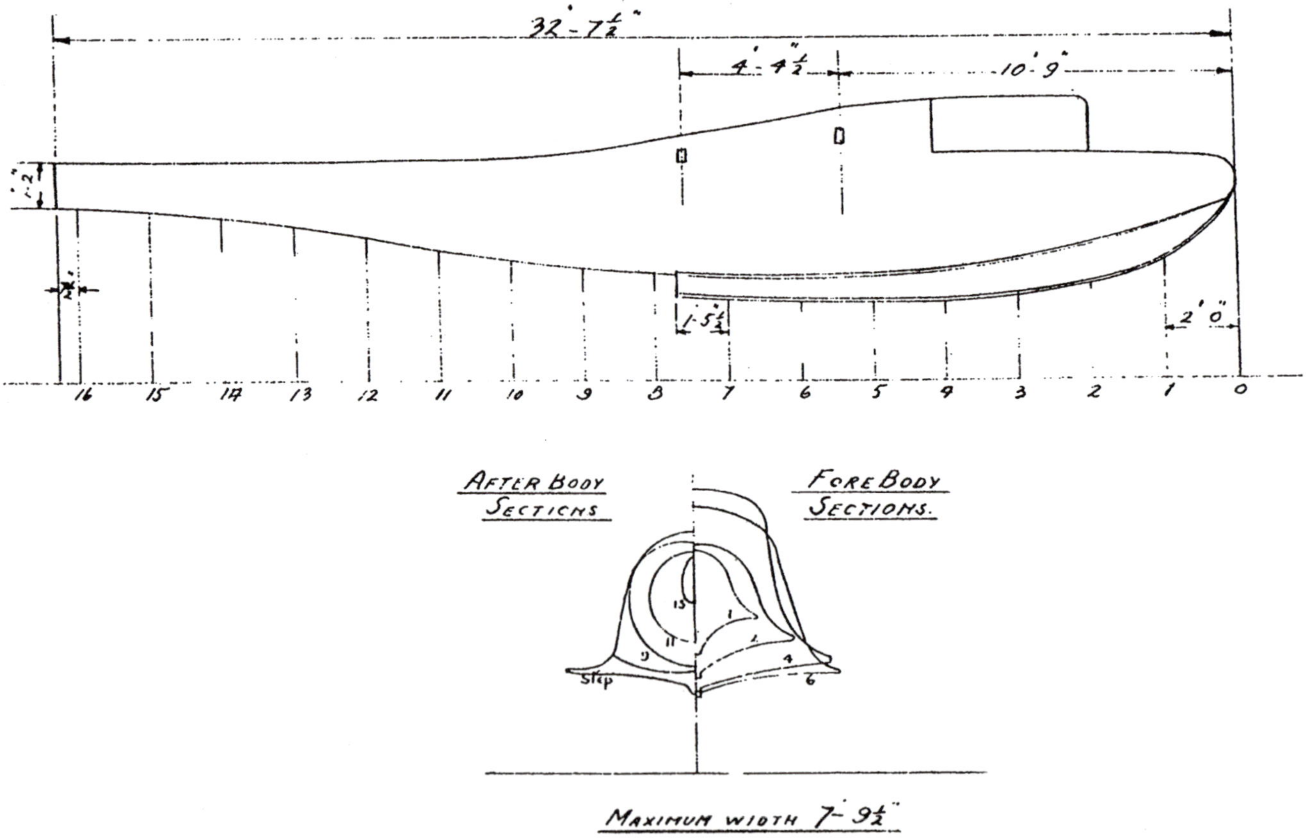

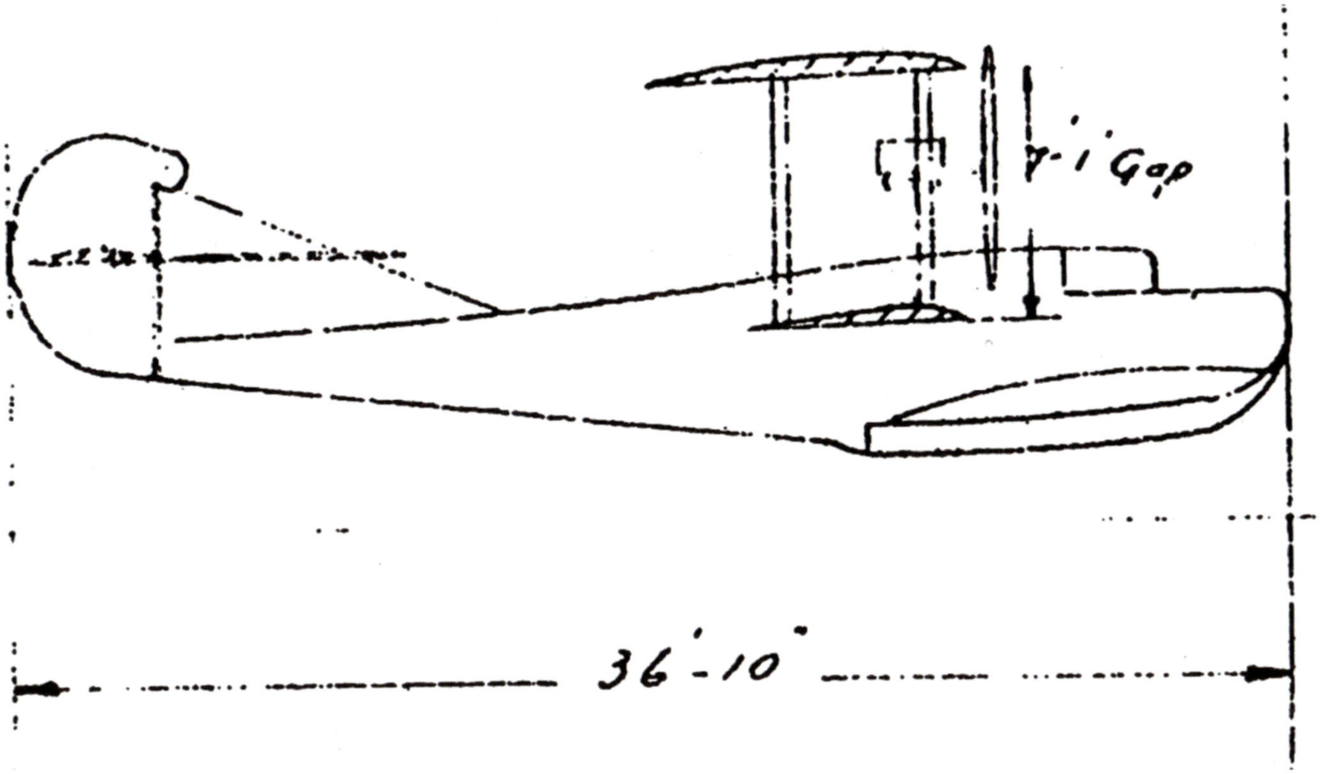
7-1 Gap
36'-10"

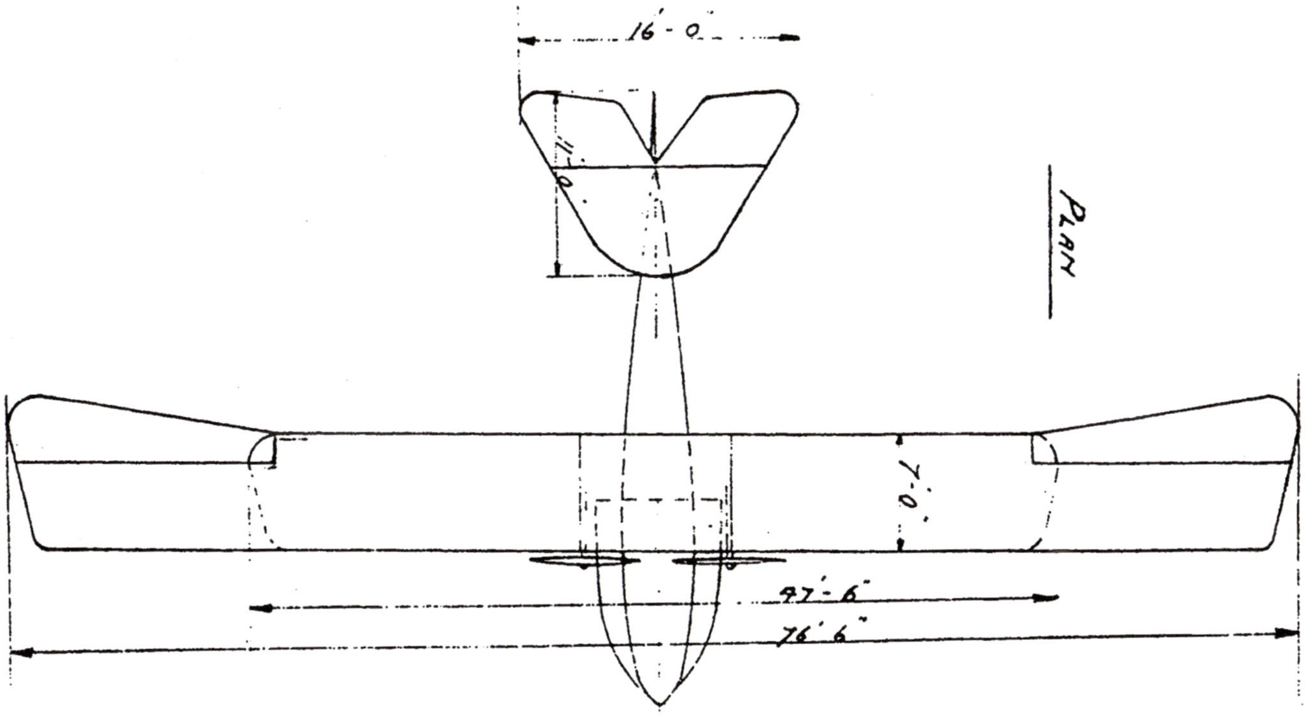
16'-0"
11'-0"
PLAN
7'-0"
47'-6"
76'-6"

Experimental Flying Boat No.3569

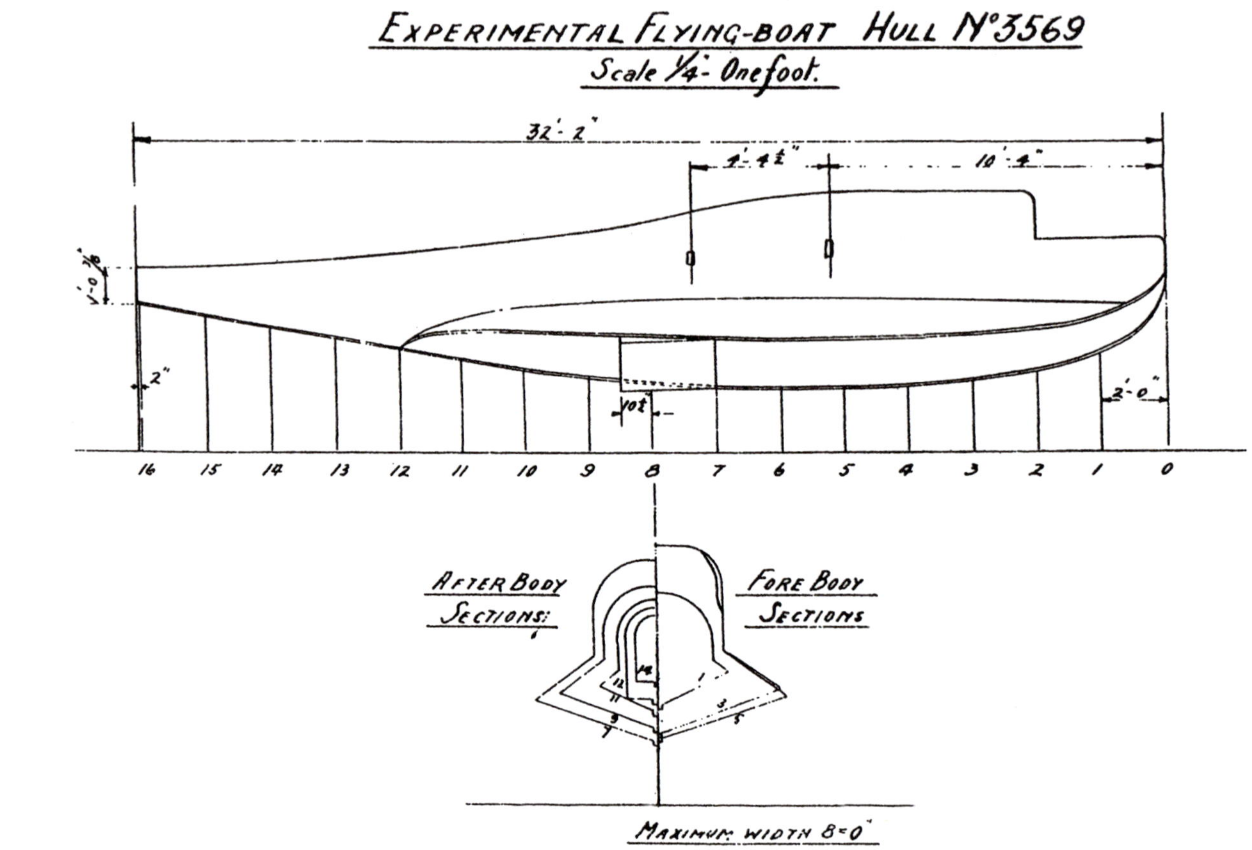

Experimental Flying Boat Hull (Porte I)

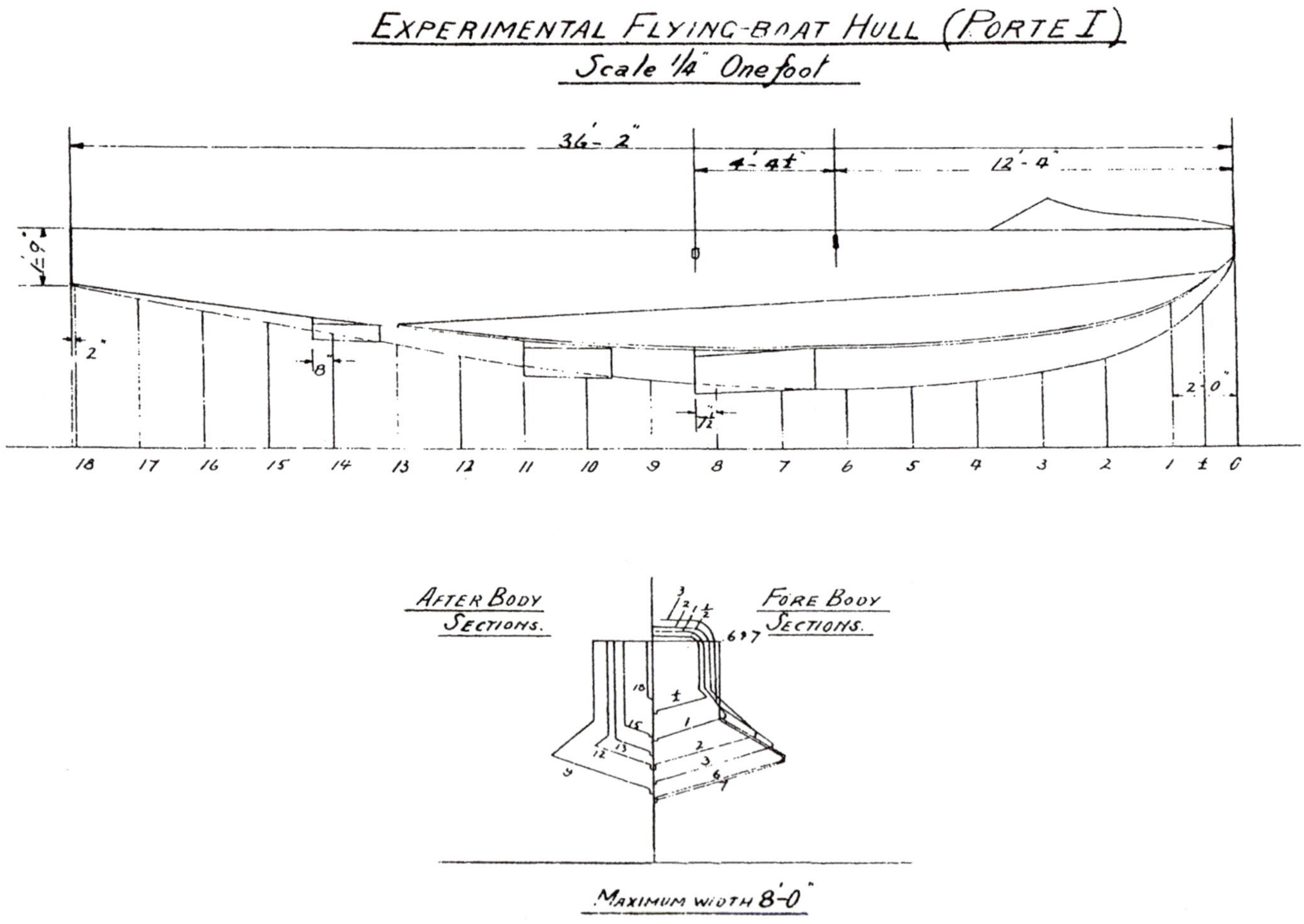

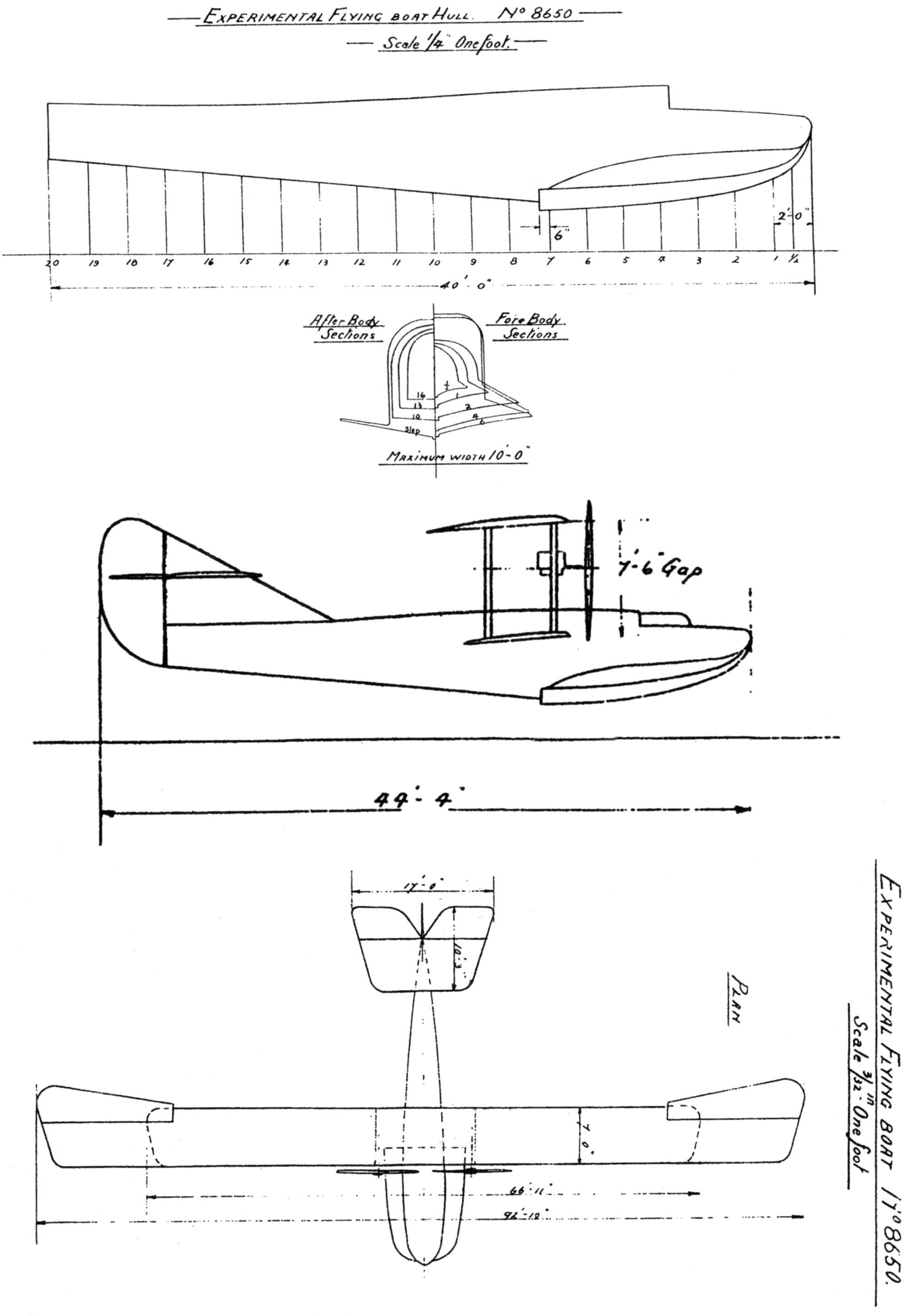
Experimental Flying boat Hull. No 8650
Scale 1/4" One foot.
2'-0"
6"
20 19 18 17 16 15 14 13 12 11 10 9 8 7 6 5 4 3 2 1 1/2
40'-0"
After Body Sections
Fore Body Sections
16
13
10
Step
1
2
4
6
Maximum width 10'-0"
7'-6" Gap
44'-4"
17'-0"
10'-3"
7'-0"
66'-11"
92'-10"
Plan
Experimental Flying Boat No 8650.
Scale 3/32" One foot

Above: N90 the prototype F.5. The F.5 was the last Porte boat to enter production, becoming the standard RAF flying boat post-Armistice.

same construction of a crossed-braced girder with simple longerons and spacers to which the fins were attached was followed. Specifications of this boat were: 16,500 lbs loaded weight; length of hull 42 ft[17]; fore and aft angle 20°. The bows were two feet longer than 8650 and owing to this increase in length it was possible to form the bows with a distinct forward flare. There were two steps, one under the rear spar and the other 7 feet aft of the spar.

This proved a much superior boat. The hump that had been so troublesome with the original hull practically disappeared, the machine accelerated evenly and rapidly to take off; there was no difficulty in getting off with 500 or 600 lbs additional load. The large fore and aft angle proved to have practically eliminated all shock on getting off and landing, and the general seaworthiness was greatly improved. In addition there was a considerable gain in buoyancy and structural strength without increase of weight. The Rolls Royce engines gave the reliability to justify long patrols. The machine retained her original serial.

Felixstowe daily reports do not record when the new hull was fitted but the machine was described as under repair from 10 to 30 June 1916. 8650 made several patrols until 30 September when she was accidently damaged near the Dutch island of Terschelling.[18] Taken in tow she was brought back close to land when the tow line parted and was lost. The wreck was washed up including the whole boat bottom that was found to be still intact. On 4 October the first Curtiss H-12 had arrived and was being erected at Felixstowe.

The Porte Baby was an attempt to build a very large three-engined patrol flying boat. Its development was separate from the evolving F-boats, and its history is related in Chapter 6.

Commodore Murray F. Sueter wrote in February 1917 that these experiments "are of great value as a ground work in flying boat research and the experience gained in these long and patiently conducted tests will, I trust, enable Wing Commander Porte and the staff of Felixstowe Air Station to continue research on hull design with success proportionate to their efforts."[19]

The Felixstowe F.2A, F.2B, F.2C, Curtiss H-16 and H-12 Converts are all grouped together as they are all variations of one model, the original F.2.

Their leading dimensions are nearly the same, and their differences are largely due to preferences of their different makers.

F.2C: N64. Built at Felixstowe, open cockpit, tandem dual control and streamline wires. Fastest boat in 1917 reaching a speed of 88 knots. Later converted to F.3.

F.2A: Built by contractors from Felixstowe drawings or modifications of them. The centre sections were wider than F.2C and F.2B to allow for larger airscrews. The early boats had cabin tops, but later ones had open cockpits.

F.2B: The name given to some spare hulls ordered to replace the *America* hulls of the H-12 machines, before completion they were altered to be interchangeable with F.2A hulls and the altered machines were known as H-12 Converts.

H-16: The name given by the Curtiss Company to the F.2A built by them.

The story of these machines is one of constant improvement and modification. The original rather crude detail design was

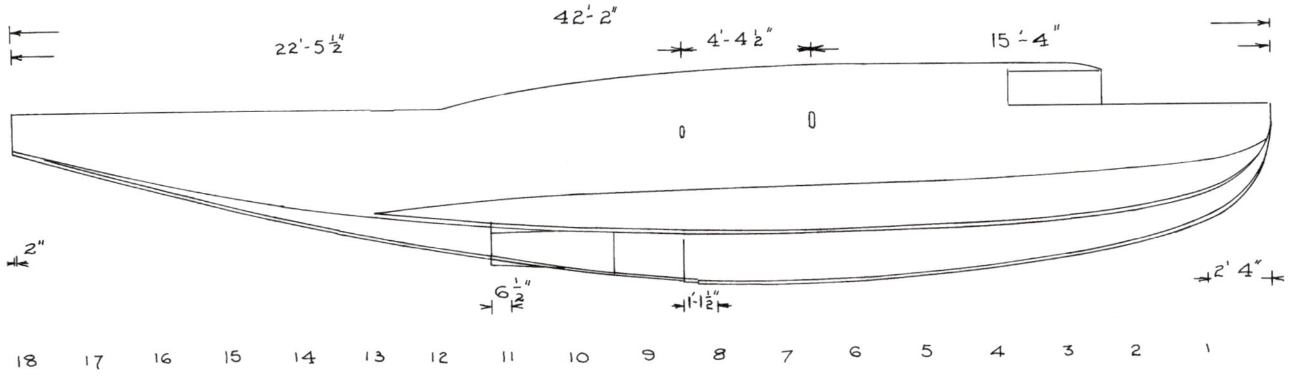

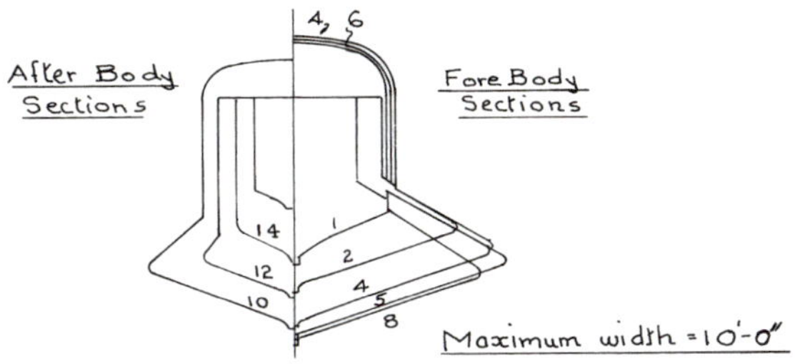

Above: Close up view of the Porte hull fitted to 8650.

Above: N4839, a production F.5, was visually similar to the prototype but the performance suffered due to official interference in the design. N4839 was from a batch of 50 boats ordered from Short Brothers Ltd. It made its first flight at Short's, Rochester, on 23 March 1920. Served with the Development Flight at Grain in 1922 and was still with the Marine Aircraft Experimental Establishment in June 1923.

perfected and petrol systems, armament, etc., were improved as experience dictated. The gross weight was increased until it had become 11,500 lb, and on special occasions, 12,000 lbs. The increase in power of the Rolls Royce engines compensated in some measure for this increase. Boats failed to get off on occasion, pounding until their bottoms failed. Top speed was usually 80 to 84 knots, while cruising speed was 60 knots.

In 1916 the demand for a long range flying boat led to the design of the F.3 to take the most powerful engines then available, the 310-hp Sunbeam. Due to the demand for this boat it was rushed through the design stage rather than incorporate many improvements in detail that experience had suggested. Redesigning would have caused delays to established manufacturing processes. The hull was still 14 feet wide, but overall length was increased to 45 feet. Gap, chord and span were increased to give an additional 300 square feet of wing area. Provision was made for 430 gallons of fuel. A speed of 77 knots was obtained on trial with an initial climb of 400 feet/minute.

The wings employed a RAF 14 aerofoil section, subsequently changed to a modified RAF 5 section. A great deal of trouble was experienced with the Sunbeam engines and Rolls Royce Eagle VIII engines were substituted.[20] The performance with the Rolls Royce engines was rather better than with the Sunbeams due to their lighter weight, and the useful load was increased.

The F.3 was officially adopted for anti-submarine work in 1917 but did not appear until a year later, but they did some good work before the end of the war. The F.3 proved to have a weakness in the design of the hull planking that would spring and leak badly. Being larger and heavier than the F.2A they were not liked by their pilots. They were restricted to areas where opposition from German aircraft was not likely.

The original F.3, the again modified N64, was completed in December 1916, and was used for patrol work during 1917 as well as participating in various trials and experiments. She outlasted many H-12 machines and was finally deleted thoroughly worn out.

It was decided to design a new boat, the F.5 (there is no record of an F.4), of the same overall dimensions as the F.3 but incorporating all the improvements that experience had suggested but had not been able to be incorporated into the F.3. Originally it was intended to fit one or possibly two Coventry Ordnance Works guns into the design but they were not perfected in time, and the machine was completed for anti-submarine work. The machine had to carry two 500-lb bombs or 4 x 250-lb bombs. The water performance of the F.2A and F.3 had deteriorated as they were more heavily loaded, so the new F.5 was made with a deeper back step and a fuller Reel line amidships. These changes were effective and the boat planned more easily than any of its predecessors.

The chief improvements besides the hull form were as

Above: The cramped conditions under which construction of Felixstowe F.3 boats took place at the Phoenix Dynamo works, Bradford.

follows:

The size and distribution of the main girder structure were revised and a stronger form of transverse bracing introduced. The junction of planking and keel was strengthened by carrying the timber right across and the disposition of the bulkheads was corrected. Any two compartments could be flooded without the boat sinking. Fuel tanks were rearranged to give a clear access along a gangway on the port side. The rudder post was strengthened considerably making the tail more secure.

A modified RAF 6 aerofoil wing section was employed. Streamline wires were adopted. The whole tail was redesigned, given more chord and lesser span, and better fastening to the hull. As the tail was so much stronger it reduced the rather alarming swaying of the tail which took place with the F.3 when the engines were run up on the ground.

The balanced ailerons and rudder overcame the heaviness that plagued the F.3's controls. A servo motor was fitted to the ailerons but hardly ever needed. An open cockpit was chosen to give a better view for alighting and fighting instead of the cabin with glass windows. The problem with the windows fogging up was thus eliminated. It was also felt that the open cockpit improved the streamline form and gave a slight increase in speed.

The Porte Super Baby (PSB) or Felixstowe Fury was designed for three 600-hp Rolls Royce Condor engines. As these engines were not available five Rolls Royce Eagle VIII engines had to be used which led to a drop in performance. The PSB was the best boat turned out at Felixstowe. (See Chapter 8)

All the F boats tend to porpoise when the boat was on the step, especially in rough seas, but there was generally sufficient elevator control to check these oscillations.

The Porte Form of Construction:

The main framework of the hull was of the box girder principle fuselage comprising two upper and lower longerons running right fore and aft meeting at the stem and sternpost. Aft of the front spar the sides and top were braced in the usual manner with struts and wire or tie rods. Forward of the front spar the sides were N girders with the top braced as above. Between and below the level of the lower longerons ran the keel and solid keelson without a break. The keel ran from the stern post

Above: The Fairey N-4 Atlanta flying boat was in development when the war ended. The N-4 program was very complicated with the work of coordinating the building of the boats with different hull designs undertaken by Fairey. The program continued post-Armistice. The first Atalanta to be completed, N119, utilised a hull built by May, Harding and May, and received the Fairey construction number F.276. This was the Dick Kerr machine. Despite being almost completed by the Armistice this machine did not fly until 1923. The photograph has been altered with rotating airscrews added to make it appear that the machine is in flight.

right around the stem to finish at the gun ring. The keelson terminated at the stem and at the stern post. Deep solid floors, on a level with the top of the lower longerons, joined up the keel, keelson and these longerons to form the backbone of the hull. The chine and fin top longitudinal sweep up into the lower longerons to complete the framework. David Nicolson was a critic of Porte and a supporter of Linton Hope.[21] He was "of the opinion that the keel is of faulty design, for many boats were found to leak badly, partly due to the bad connections between the keel and bottom planking, and partly because the keel is too narrow. The keel and planking are fastened with only one row of brass screws which secure the bottom planking to the keel in the hulls of the F.3 type."[22]

In the F.5 the timbers were continuous from fin chine to fin chine, notched through the keelson on keel level and from chine to fin top longitudinal, using copper rivets, forming a much stronger combination than in previous F boats.

A timber was fastened on each side of a floor, of which there was usually one at each strut position and one between struts, and two timbers between each floor. Planking was fastened to these timbers. Several fore and aft stringers were fitted, notched out to receive the timbers to which they were fastened. On the original F boats the planking was double diagonal, continuous from stem to stern, fore and aft, and to chine athwart ship, and the step planking added separately. The fin tops and sides to just aft of the rear spar were planked with three-ply, aft of which the sides were covered with doped fabric, a solid mahogany washboard about 1 foot deep extending along this length. The top from aft the pilot's cockpit was covered in doped fabric laid on fore and aft stringers, supported on formers, forward of which the top of the hull was planked.

Bottom planking was arranged on a diagonal system the inner skin of cedar 1/8 inch thick at the ends and 3/16 inch thick amidships, fitted at an angle of 45° inclination to the keel. The outer mahogany skin was 5/32 inch thick forward, 3/16 inch thick amidships and 1/8 inch aft. The planking being laid at an angle of 30° with the forward end of the planks butting the keel. A layer of varnished fabric was fitted between the two layers making a strong structure.

The fin top on the first series of F boats was of three-ply birch, and in later types was covered with fabric and varnished. The timbers under the fin of the early F.3 hulls were heavy and widely spaced. On later boats smaller timbers spaced closer were substituted to permit all though fastening of the diagonal planking on the hard wood timbers. All the fins on the F.3 were flat, but on the F.5 were given a ½ inch camber, that added strength, and assisted in getting rid of water easily.

The timbering at the bow was composed of rock elm and reinforced by horizontal stiffeners below the top longerons. Above the top longerons there were 10 deck-stringers that were notched to take the ribs, together with three strong beams which subdivide the athwartship ribs. This skeleton, which was shaped like a dome, was planked diagonally, the inner skin being laid at 45°, with the outer skin being laid approximately fore and aft to suit the curve of the nose. Each skin was of mahogany 5/32 inch thick, the planking being fastened to the strong beams with wood screws. Although slightly heavier than the rest of the hull, "it is the best piece of construction in the whole boat."[23] Strength of the fore body is essential for at high speeds the resistance of the air is great and a weak nose would be very easily damaged and driven in.

Abaft the bow planking the sides were of three-ply birch

and extended from the bottom of the fin member to the top longeron, running aft to the gun-port openings, a distance of about 18 feet. The rear of the hull had fabric sides, however there were incidents where the fabric did not stand up to the sea and boats were lost. This happened with RAF F.2A and USN H-16 boats. The fabric was eventually replaced with diagonal planking of two skins, each 1/16 inch thick with nainsook between. This was a great improvement as it increased torsional strength and was only about 47 lbs heavier.

For operation from sheltered harbours such as Felixstowe the hulls were suitable. When operational requirements meant that they had to be carried out under less favourable sea conditions, several weaknesses became apparent. The joints leaked, the three ply on the fin tops and hull sides rotting and opening up the laminations. Similarly wash boards split and the fabric rotted. The fin tops were then double planked diagonal with mahogany and cedar, and the sides were either single planked fore and aft, and fabric covered, or planked with "Consuta."[24]

The solid transverse mahogany floors were fastened to the lower longerons by metal angle plates and notched for two thirds of their depth for the bottom to fit over the keelson. The keelson was notched out one-third of its depth from the top. Although the keelson was nearly 12 deep in parts it was greatly weakened by having one-third of its depth cut away to accommodate the floors. The corner of the joint were filled with square fillets for the whole depth. This was a weak and uneconomical form of construction, resulting in frequent splitting of the floors. The floor was cutaway about two-thirds its depth, thereby sacrificing strength to accommodate the keelson. In Nicolson's opinion, the "keel is of faulty design ... for many boats were found to leak badly, partly due to the bad connections between the keel and bottom planking, and partly because the keel is too narrow...only one row of brass screws secure the bottom planking to the keel in the hulls of the F.3 type."[25]

The steps on the F.2A and F.3 were framed with ash bearers ½ inch thick, and were three inches deep at the after edge, tapering to meet a board that ran off to a feather edge forward. The steps were of the open type and the planking was fastened with wood screws to the ash bearers. The inner skin was cedar with a mahogany outer skin, both laid diagonally. The whole step was constructed on a bench and then screwed onto the bearers. In some cases the steps would be wrenched off the bottoms of the hull just prior to the boat getting off the water. This was remedied in the F.5 by carrying the inner skin of the bottom right through from end to end of the boat. The outer skin was then put on and carried forward to a feather edge under the main step.

Nicolson considered that another "weak point (in the F boats) is the discontinuity of transverse strength caused by running the timbers down to the keel and stopping them there, no provision really being made to hold the centre girder to the bottom planking or sides of the hull."[26]

The hull framework, wings and tail structure formed a complete and simple braced structure independent of the hull skin. The only drawback to the wing root structure was the absence of a clear passage through the hull. The wing root spars were one of the most important structures in the machine. They not only carried the weight of the wings, but they supported the engines. When in the air the hull was suspended from them. They were held in position at the sides and centre of the hull by heavy stanchions and struts. They were originally solid wood but in later boats were laminated in two or three sections. Outside the hull they were covered with three-ply birch.

The top of the hull had many openings. The foremost being the gunner's cockpit with gun ring, followed by the pilot's cockpit. Further aft was the wireless operator's hatch. The engineer's hatch was placed on the port side immediately aft of which was a three-ply footway extending across the hull. Another opening accommodated the wireless mast and lastly there was an 8 inch triangular ventilator. The deck aft was built up with the minimum number of stringers and covered in fabric in order to keep the weight to a minimum. The heavier the rear of the hull the greater the bending moment and tendency to shear. The interior of the hull was well lighted due to the doped fabric top.

The front gunner's cockpit had a table that extended from his seat to the nose of the hull. Underneath the table was an ammunition box for the Lewis gun trays. The three-ply hinged seat had a high back. The pilot's cockpit's seats for the pilot and assistant pilot had well upholstered seats of kapok cushions that could act as flotation devices if required. The assistant pilot's seat folded out of the way to open up a clear passage aft. A few feet behind the pilot was located the wireless cabinet together with the operator's seat, while on the opposite side a ration box was fitted. The engineer's compartment was situated aft with a ladder giving access to the top of the deck.

A considerable number of metal fittings were used in the construction of the F boats. The chief elements of the structure, such as struts, floors, wing root spars, etc, were held together by light steel fittings. The fittings in the F boats were so many that their manufacture in quantity had a considered influence on production of the boats during the war. Despite Major Rennie's assertion that they "are readily inspected and do not corrode or rust away in a few weeks as some critics would have us believe. With little attention they will outlive the hull," many of these were in inaccessible places and to replace them would mean dismantling the hull.[27]

One criticism of this type of construction was that it did not readily admit to the fitting of a double bottom. Major Rennie answered that the F boats on operations had shown that damage caused by flotsam was a very infrequent occurrence. While Capt David Nicolson agreed that the F boats were a "great improvement on the Curtiss type, and were fitted with engines of considerable greater power," he also noted that "as in previous boats, the bottom gave trouble owing to faulty floor and keel construction."[28]

Comparison of Felixstowe & Hope Hull Designs

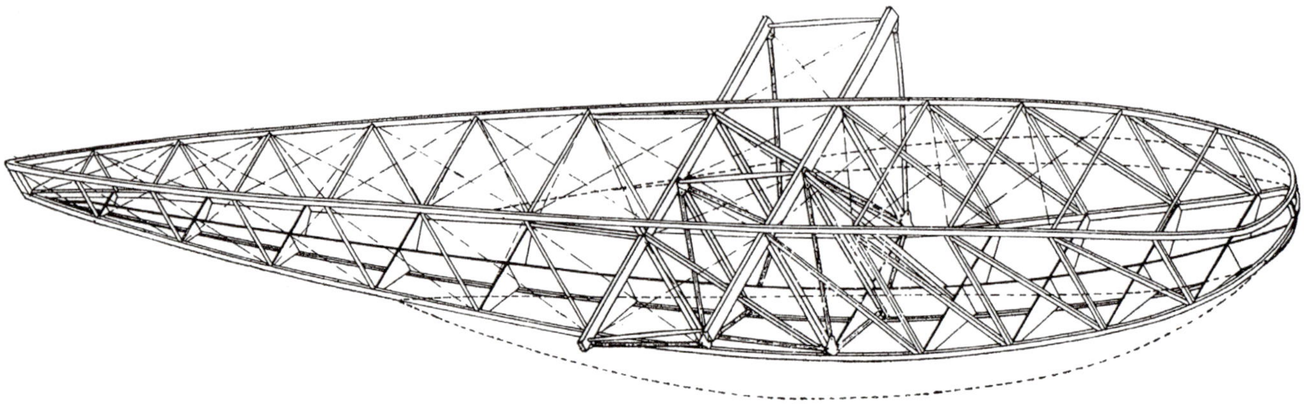

Fig. 7.

Details of Felixstowe Hull Design

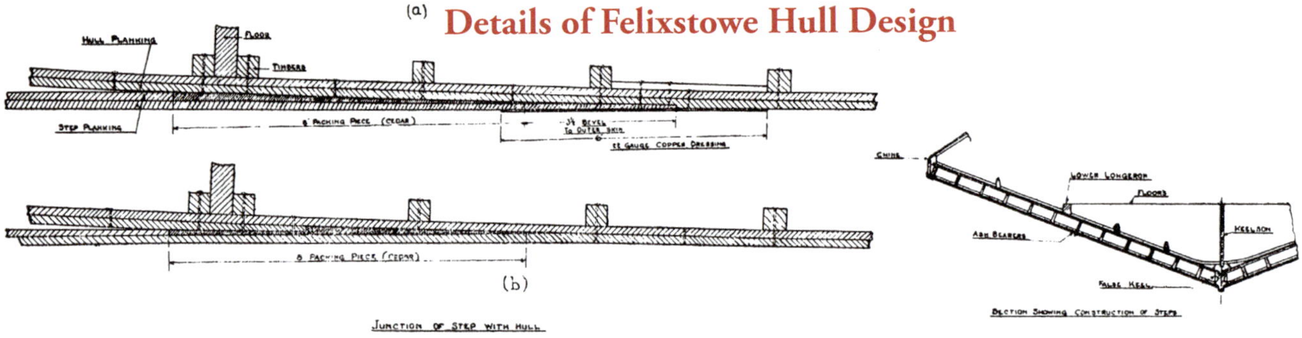

Details of L. Hope Hull Design

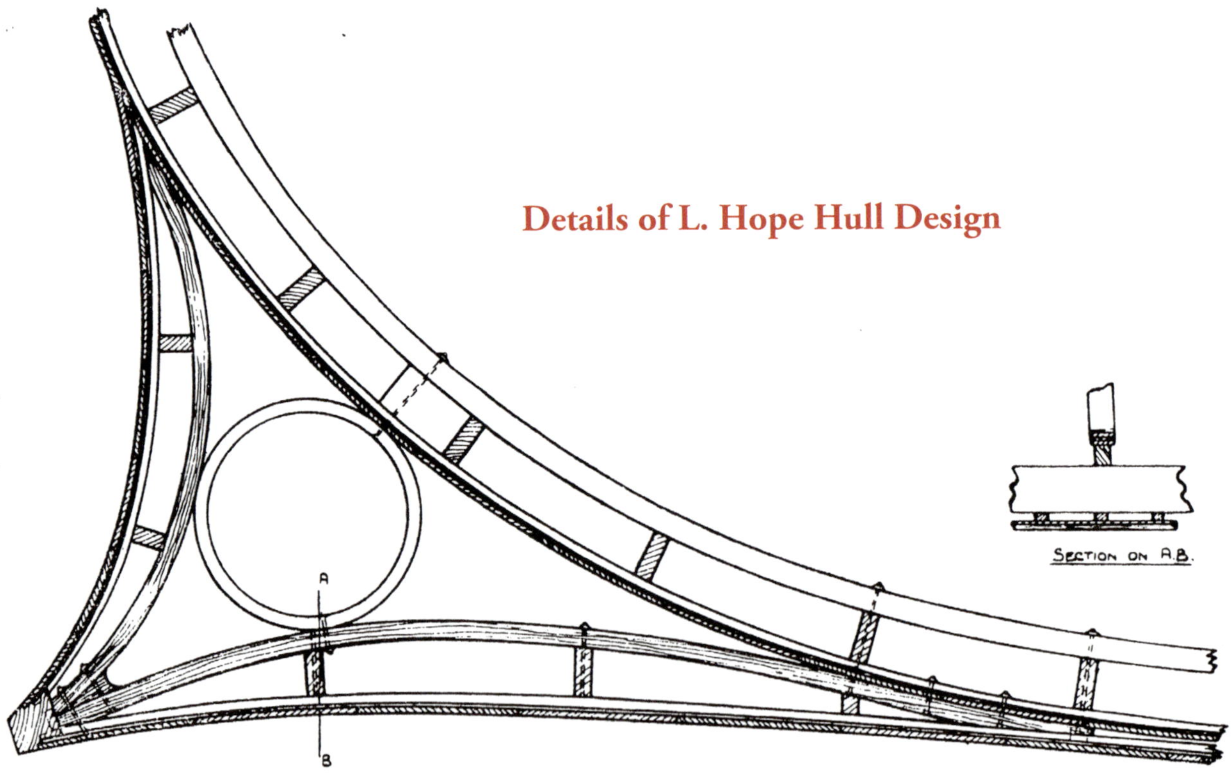

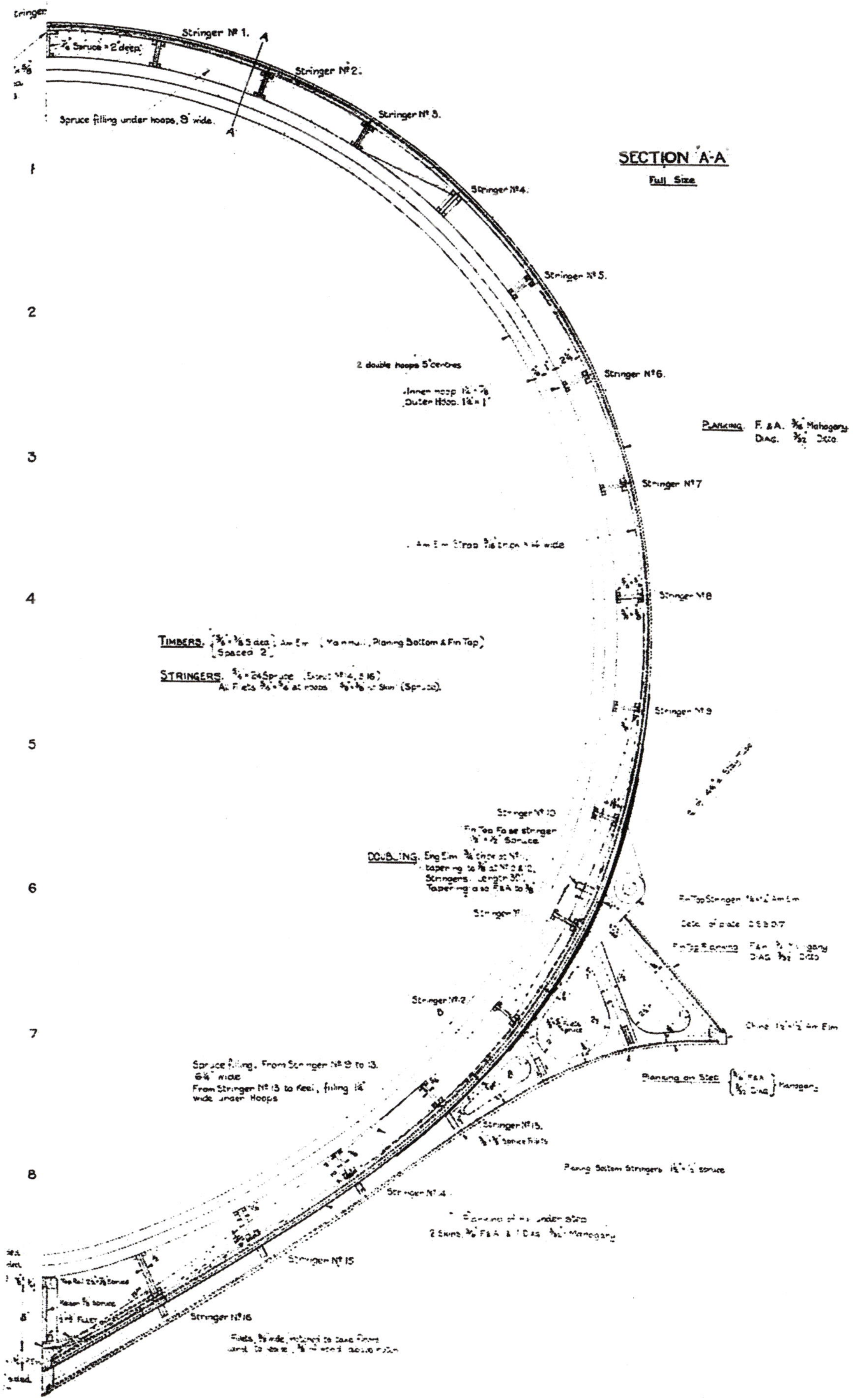
Stringer Nº 1.
7/8 Spruce × 2 deep
Spruce filling under hoops, 9 wide.
Stringer Nº 2.
Stringer Nº 3.
SECTION A-A
Full Size
Stringer Nº 4.
Stringer Nº 5.
2 double hoops 5 centres
Stringer Nº 6.
Planking. F. & A. 3/16 Mahogany.
Diag. 3/32 Ditto.
Stringer Nº 7
Stringer Nº 8
Timbers.
Stringers.
Stringer Nº 9
Stringer Nº 10
Doubling.
Stringer Nº 12
Spruce filling. From Stringer Nº 9 to 13.
6¼ wide
From Stringer Nº 13 to Keel, filling 1¼ wide under Hoops
Planking on Step
Stringer Nº 13.
Planing Bottom Stringers
Stringer Nº 14
Stringer Nº 15
Stringer Nº 16

Comparison of the Porte – Linton Hope Methods of Construction

The Linton Hope or flexible type of construction was totally different to the Porte type. It consisted of a keel and keelson continuous from stem to stern post, and a large number of fore and aft stringers distributed evenly around the periphery of the hull that was of streamline form Small section timbers closely pitched were bent round the stringers, ending at the keel.
To these timbers was attached the planking, the inner being diagonally laid, and the outer fore and aft. The planning surface extended from the bows to the main step and was attached to the three-ply formers on the hull bottom and at the step, which was closed, to fore and aft bearers. The fin top was straight and the chine ended at the step. The after step was attached in a similar way. This form of construction led to a double bottom in way of fins.

Nicolson promoted the Linton Hope hull. Linton Hope was a naval architect and designer of racing yachts. The Linton Hope form of construction was entirely different to the method Porte developed.[29] This was claimed to offer fair and easy lines from a circular cross section, consequently offering less air resistance and higher speeds for the same horsepower. They were stated to be generally stronger weight for weight than the F boats, and more seaworthy. The F boat was said to be a compromise with a flying boat forebody attached to a fuselage tail. According to Nielson this compromise proved weak and, as described above, additional planking had to be added to strengthen the hull. Having rectangular hulls they were stated to be weaker in transverse strength than the Linton Hope circular design.

Linton Hope designed the hulls of the A.D., the Phoenix P.5, and the Fairey N.4 flying boats. The A.D. boat was not a success, and while the P.5 had potential the Armistice ended any chance of it being produced in quantity. Nicolson considered the P.5 as "being far ahead of anything previously accomplished."[30] It was stated that the complicated movement of the elevator to get an F boat off the water was not required with the Linton Hope type as these were designed from model results in the tanks of the National Physics Laboratory, and none of these control movements were necessary. Also the Linton Hope type did not need the complicated wing root structure necessary in the F boats.[31] It was noted that the flying boats were hard to hanger and were often left out at moorings. The Linton Hope hull was considered by Nicolson to have better qualities when moored out.

Replying to Nicolson's 1919 paper, John Porte pointed out that the F boats had a long war record no Linton Hope hull had seen war service and had only been in use for a limited time as experimental models. Nicolson had made no attempt to prove that the Linton Hope type of hull was stronger weight for weight than the F type and as the Linton Hope boat had not been subject to rough usage and handling under actual service conditions, "it did not seem justifiable to make such a statement...During a period of 18 months two seaplane stations operating F boats on the east coast on submarine and reconnaissance patrols had 230,000 sea miles flown to their credit, and not one man lost through unseaworthiness of these boats, although several forced landings had to be made in the North Sea either due to enemy action or failure in the power plant."[32]

The size of the boats meant that they had to be housed in sheds except where they operated from sheltered harbours and even there they had to be frequently taken out of the water for repairs and overhaul.

To bring in the boats they were put on trolleys while in the water. The trolley ran up an inclined slipway to suit the rise of tide, and were pulled up with the boats in place, and housed in a shed. Now, experience has shown that, apart from crashes and really bad landings, the most serious local stresses which a hull had to withstand were those suffered during its life on the trolley, or getting off and on the trolley. If a flying boat were to be safely and quickly put on a trolley, and, when on, not be distorted locally, the whole weight should be taken on the trolley at the keel. With the F type that could be done, as the transverse bracing at the centre section, where the wing structure joined the hull, was so designed that the resultant loads due to the various weights, such as wings, engines., passed through the keel. The necessary side support was taken off the side bottom longerons, which also formed part of the transverse structure, but no load was taken by the fin or hull planking between the keel and longerons. In the Linton Hope type that was not possible, as there was no such transverse bracing, the result being that unless special props were provided on the trolley to take the load under the engine struts, the hull would distort locally, probably to such an extent that it would be very difficult to keep the hull in the vicinity of the step watertight. Time and experience alone could tell which type of construction would survive as being most suitable to fulfil the many and complex conditions of service.[33]

The critic of Porte hulls, David Nicolson, considered that the F boats were too nose heavy due to the stringers being too closely spaced, while the timbers and bottom were much under strength.[34] Nicolson had "pointed out for many years that any hull construction methods employed on the F.2A. boats could not be seaworthy." He considered that the hull of a flying boat should be as seaworthy as a motor boat. "A seaworthy hull that could fly was what was required, and could be accomplished if the best yacht builders were given a free hand."[35] G.S. Baker on the other hand stated that the "real defect" of both the F and P type hulls, that is Porte and Linton Hope, hulls, was that "one could not get at them inside." For impact experiments on an F hull he had been given a hull "not quite three years old, and the inner skin was rotten in places, not because of bad work, but because it was almost impossible to get there to keep it clean. One could hardly get at some of the structure inside the boats, and if the hulls were to last more than two or three years they must be readily accessible inside."[36]

"The fact remains that the Porte hulls stood up and are still standing up to the work for which they were designed, in spite of all the adverse criticism and condemnation which they have been and are subject to. Admittedly there are many faults and weaknesses in the Porte hulls, but there are just about as many

Above: The A.D. flying boat had a Linton Hope designed hull. Supermarine "built" the type under Contract AS5388/17, (N1520-N1529). N1525 bears Supermarine's logo on the anti-skid fins that were fitted between the interplane struts.

Above: The first Phoenix P.5 prototype N.86 prior to first launch. The smooth streamlines of the Linton Hope hull are well illustrated. (For the story of the P.5 see *Cross & Cockade International* Vol.45/1 P.3.)

in the Linton-Hope hulls."[37] What must be remembered was that Porte was not a theoretician but flew his boats in combat and knew what was necessary to produce a fighting machine that could be produced by firms without the boat building skills needed for the Linton-Hope form of construction.

Just before the Armistice work was in hand to fit steps to the F.5 in accordance with results developed in the NPL tank. This work ceased due to demobilisation. Porte developed the position of the steps from observation as the result of a great deal of experience of taking off and landing flying boats in all types of seas and weather. The experimental F.5 had been flown before there were any full-scale tests of the tank arrangement of the steps.

The US Navy used the F-5-L successfully for many years post war. Maj Buchanan "did not wish to diminish what had been done by the late Major Linton-Hope, with whom he had been in close contact at the Air Ministry and the Admiralty, he pointed out that while he agreed with the criticisms of the F type boats, he thought that to claim that the Linton-Hope method was a complete solution was "going rather long way." His reason was that there was a lot of experience that had been gained with the F boats while that with the Linton Hope hull was very limited.[38] The wooden hulled Supermarine Southampton utilised a Linton Hope type of hull, but this was soon superceded by the metal hulled Southampton. The argument as to which hull construction was the better was now of academic interest only.

Post-war Porte made a claim to the Royal Commission on Awards to Inventors for the work he had carried out on developing the flying boat during the war. The claim was separated into two separate cases:

1. The claim in respect of the American Flying Boats delivered in Felixstowe in August 1914.

2. The claim with respect to the subsequent modification of this flying boat.

The claim was considered under three heads:

(a). What was the value of the novel features in this boat?

(b). How far were these novel features due to Colonel Porte?

(c). How far had Colonel Porte parted with any rights he may have had?

It was considered that the novel features in (a) were not of great inventive ingenuity, but were a natural development of the flying boat. It was acknowledged that development of the first America boats was done in the USA and it would be most difficult to obtain information as to the design sequence. As to part (c), it appeared "that Colonel Porte had certainly parted with some rights to Wanamaker, and it was through the letter that the Aircraft Manufacturing Company was placed in possession of the material necessary to enable them to construct the original boats." This is a reference to the eight H-4 flying boats built by the AMC in 1914-1915. Wannamaker had made a claim against AMC and "the indications point to some probability of this being eventually borne by the Department."[39]

In the summer of 1923 the US sent a Commission[40] to London where it met with a British Commission to settle the claims of inventors for the use of their inventions during the war. For use during the war years the compensation was paid by the British Government as per the Bolling Agreement. Royalties for use post-war were paid by the US Government. The principal claim that concerned the USN was that of the estate of Commander Porte. It was held that Porte's work was a natural development of the flying boat, the basic patents of which were held by Glenn Curtiss. Moreover as Porte was a serving officer at the time he had done his work in the line of duty. A nominal sum of £1,500 was paid to Mrs Porte as a gratuity for his services to the USN.[41]

Capt A.E. Bolton noted in his paper on the development of the large flying boat that the "advent of the Large Flying Boat marked the beginning of a new era in the naval side of aerial warfare, as they were a very distinct improvement over the large but slow float type sea-planes at that time in use in both air speed, observational powers, bomb carrying and radius of action."[42] It was the Porte boat that took the action to the enemy and this is his legacy.

F.2A assigned to Felixstowe NAS in dazzle scheme.

Chapter 2 Endnotes

1. First Statement made 1st August, 1917, Commander John Cyril Porte. Copy in The Royal Commission of Awards to Inventors, Claim No.580. Claim of Mrs Mynne Porte. Copy supplied by late Ray Sturtivant. TNA WO158/993.
2. For example, the Curtiss HS-1L/HS-2L series had ailerons on all wings and those on the lower wing were continually damaged by the sea or by handling when bringing the machines ashore. Boeing developed a set of wings for the HS-2L that had larger ailerons only on the upper wing.
3. Rennie, J.D. "Some Notes on the Design, Construction and Operation of Flying Boats," *The Journal of the Royal Aeronautical Society*, Proceedings, Seventh Meeting, 58th Session. 18.01.1923. P.123.
4. Baker, G.S. & Keary, E.M. in conjunction with Capt Gundry & Lt Hackforth of the Isle of Grain Experimental Station. *Experiments With Full Sized Machine. First Series.* Advisory Committee for Aeronautics, Reports and Memoranda No.473, September 1918. P.3. The RAF C.E.I was finally used for the purpose of full sized experiments. Baker was to become a supporter of the Linton Hope form of construction. See Rennie, J.D. *Op. Cit.* P.108.
5. From Nicolson, D. "Glossary of Terms used in Flying Boat Hull Construction", published by NACA, Technical Memorandum No.40. Taken from *The Aeronautical Journal*, August 1921.
6. It appears from Curtiss testimony to the Royal Commission for Awards to Inventors in the post-war years that these first boats were designated H-2 by Curtiss.
7. Sturtivant, R. and Page, G. *Royal Navy Aircraft Serials and Units 1911–1919. Op Cit.* P.67.
8. The initial order included 166 *America* type flying boats, Porte was sent to the US to sort out this and other contracts with the Curtiss Co. and the H-8, which Porte claimed to have designed, was ordered as a replacement to complete the order.
9. First Statement made 1st August, 1917, Commander John Cyril Porte. *Op Cit.*
10. The Commission used this direction to state that all rights associated with Porte's experimental work were the Government's and reject Mrs Porte's claim. "Before the Royal Commission on Awards to Inventors – In the Matter of a Claim by Mrs Mynnie Porte in respect of "Flying Boats.", Date stamped Jun 1923. Material supplied by Ray Sturtivant from UK National Archives.
11. Sueter, M.F. *Report on Experiments with Seaplane Hulls, carried out at RNAS Air Station, Felixstowe, during 1915–196 – Presented by Commodore Murray F Sueter, C.B., R.N., Superintendent of Aircraft Construction, Admiralty, to Advisory Committee for Aeronautics.* 01.02.1917. TNA DS.IR23/917.
12. Sueter gives this change in dimensions. Rennie gives "Hull 2 ft. longer in nose and tail than No. 3545, being 36 ft. overall." Both give the same hull length but the plan of the Porte I shows an overall hull length of 36 ft 2 in.
13. What does one have to do in order to be legally known as the designer of an aircraft? Porte did not have formal engineering training but would have been up to date with the technology of his era as he was intimately involved in submarines. According to Maj J.D. Rennie, Porte would lay down the broad outlines of what he wanted and the draughtsmen at Felixstowe produced some drawings reflecting this, however a good deal was left to the boat builders to work out things as they went wrong.

    The original Curtiss wing structure was modified quite a bit for the F-boats. At one stage the wing section was altered to facilitate production by other contractors but this was detrimental to flying.

    Porte did not talk much due to his TB and he had to be obeyed the first time. Notes of Interview with Maj J.D. Rennie, 15.09.1964. Copy in RAFM JMB Collection Box 40.
14. See Owers, C.A. "Curtiss Canada". *Cross & Cockade International*, Vol.42 No.4, 2011, P.227.
15. See Owers, C.A. *Curtiss H-12*, Datafile No. 125, Albatros Publications, UK, 2007.
16. Rennie gives 11 ft as maximum beam.
17. Rennie gives the hull length as 56 ft 10 in.
18. Sturtivant & Page, *Op. Cit.* give the date as 30 September 1916.
19. This section has been compiled in the main from M.F. Sueter's Presentation to the Advisory Committee for Aeronautics, and Major J.D. Rennie's address to the RAeS. All quotes are from Sueter's presentation unless otherwise noted. It is obvious that this address was the basis for Rennie's paper. The two documents differ in some details as noted. Maj W.F. Vernon's Report, (revised by Col Porte), entitled *Flying Boat Development*, (TNA MUN8/35), gave a historical overview of the experiments at Felixstowe and provided much of the detail material for all the boats up to and including the Fury. This Report was stated as being aimed at promoting the use of the civil use of the flying boat.
20. The first Eagle VIII engines were to be despatched to Felixstowe for installation and test on 27.08.1917. 28th Weekly Report of the Controller, Technical Department for Week Ending 29.08.1917, P.5. Copy in RAF Museum J.M. Bruce Collection Box 59.
21. Capt David Nicolson was a naval architect and marine engineer. He served in the Highland Light Infantry in France and was wounded. He returned to the UK in 1915 and joined the RNAS. He worked with Porte at Felixstowe on the development of the F boats. After a time he was went to London HQ where he was in charge of production and inspection of flying boats. While attached to the inspection and production staff, he was sent touring round several shipyards inspecting F.2A, F.3 and N.T.2B boats. He could not believe that these boats had been designed

by naval architects and was relieved when he found out that the boats "had not been designed by shipbuilders." Nicolson claimed that the improvements to the F-boats came from the Air Ministry, and not due to Felixstowe at all, several having been designed by himself. He was a supporter of Linton Hope, another naval architect, and not of Porte as can be seen from his writings. At the end of the war he returned to the Air Ministry as chief production officer in charge of flying boats. In 1923 he left to become a member of the board of S.E. Saunders Ltd. *The Journal of the Royal Aeronautical Society*, August 1920, P.464, and Flight, 23.01.1923 P.55 and also 24.06.1929. P.66.

22. Nicolson, D. "Flying Boat Construction," Paper read to the RAeS, reproduced in *Flight*, 31.03.1921. P.228.
23. Nicolson, D. "Design and Construction of Flying Boats", paper read at before the Institute of Engineers and Shipbuilders in Scotland, reproduced in *Flight* magazine, 03.07.1919. P.875.
24. This method of construction comprised the cementing together of several thicknesses of wood with grains at angle layer to layer. The resulting ply-wood is then sewn with copper wire on a special machine, the stitching giving increased strength to the resultant product.
25. Nicolson, D. "Design and Construction of Flying Boats", *Op. Cit.*, P.874.
26. *Ibid.*
27. Rennie, J.D. *Op Cit.* P.123.
28. *Ibid.* The problem with flying boat hulls was known as a letter from Col W. Sempill, Ass Controller, Technical Dept, details: All the "trouble is due to lack of sufficient transverse strengthening members and incorrect method of fitting the planks on the hull bottom." Sempill went on to say that modifications to correct these defects had been made 3 months ago; however, due to the number of hulls in store, and the lack of trained personnel, the work was proceeding slowly. "Arrangements have now been made to supply a number of Shipwrights and super-structure Builders and R.A.F. Stations, in order that this work may be carried out." Sempill to Brig Gen Groves, Director of Air Division. 15.10.1918. TNA MUN8/28.
29. Linton Hope was outspoken in his criticism of Porte and considered him a "complete amateur" both as a boat builder and a naval architect. Penrose, H. *British Aviation, Great War and Armistice*, Putnam, UK, 1969. P.61.
30. Nicolson, D. "Design and Construction of Flying Boats," *Op. Cit.* P.874.
31. Rennie, J.D. *Op. Cit.* P.123.
32. Nicolson, D. "Design and Construction of Flying Boats," *Op. Cit.* P.317. It is interesting to note that *Flight* did not publish the discussion to Nicolson's paper, thereby allowing Nicolson's statements, as printed, to go unchallenged .
33. J. Porte's reply to Nicolson's Paper "Design and Construction of Flying Boats." *Op. Cit.* P.320.
34. Nicolson, D. "Flying Boat Construction," Paper read to the RAeS, reprinted in *Flight* 24.03.1921. P.213.
35. Air Commodore C.R. Samson, "The Operation of Flying Boats in the Mediterranean." *Royal Aeronautical Society Journal Proceedings* 60th session, 05.02.1925. P.127.
36. Rennie, J.D. *Op. Cit.* P.169.
37. *Ibid.*
38. *Ibid.* P.173. Buchanan was head of the Aircraft Design Dept, Directorate of Research, Air Ministry, and was closely in touch with all the latest developments in aircraft and had an accurate knowledge of the work done in the development of the Felixstowe boats. Rennie commented that as he was also an engineer and not a naval architect, "his opinions may be looked upon as unbiased."
39. Opinion: *Claim by Lt Col J.C. Porte*, W. Trevor Watson, 26.03.1920. TNA AM211 TS28/17.
40. The US Commission members were Col MacMullin of the War Department; Mr Knight of the Department of Justice, and Cmdr H.C. Richardson of the Navy Department.
41. Cmdr J.C. Hunsaker. *Notes on the Development of Naval Aircraft. Origin of Naval Aircraft Building Program; Joint Army and Navy Technical Board on Aircraft; (Board on Design and Specifictions for Aircraft); Bolling Commission.* Copy in Hunsaker Collection, MIT Institute Archives. Copy via G. Rossano.
42. Bolton, A.E. *The Development of the Large Flying Boat for Naval war purposes 1917–1919*, TNA Air 1/2397/267/5.

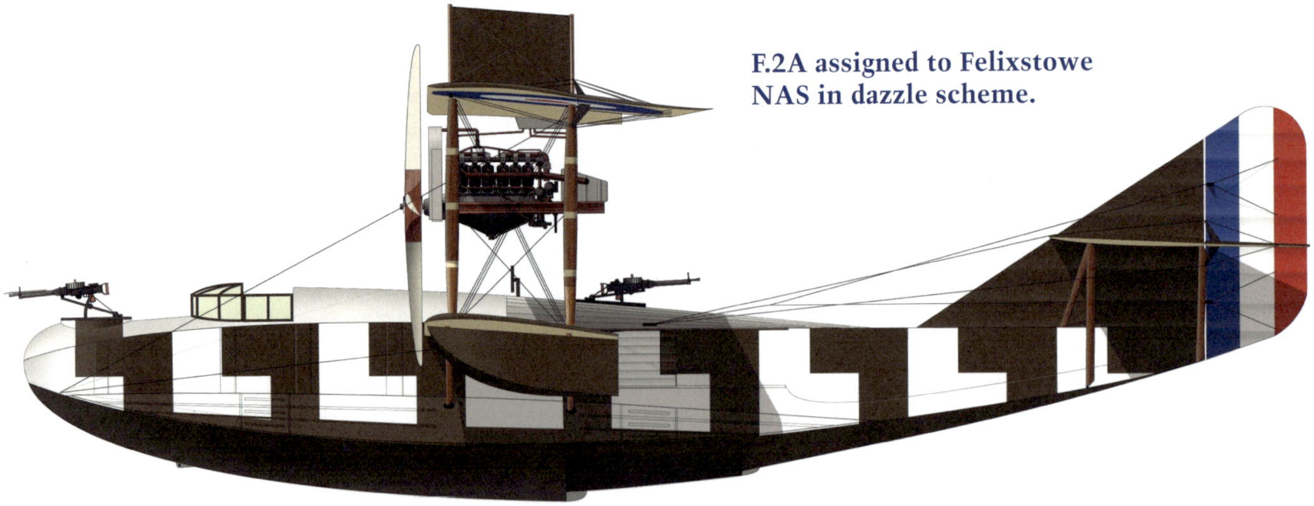

F.2A assigned to Felixstowe NAS in dazzle scheme.

Above: Another view of the cramped conditions under which construction of Felixstowe F.3 boats took place at the Phoenix Dynamo works, Bradford.

Below: View of a damaged F.3 boat showing the front spar connection to the hull.

Above: View inside the front gunner's cockpit looking forward to the bow.

Below: View of the same damaged hull shown at bottom of preceding page with the wing section at the hull removed.

Above: F.3 hull with the ply sides added but the fins not yet planked.

Below: Note the stencilling on this F.3 hull. A plumb bob is used to align the parts of the tailplane with the hull.

Above: Detail of the upper wing gravity tank piping.

Below: Wing joint with aileron control cables along leading edge.

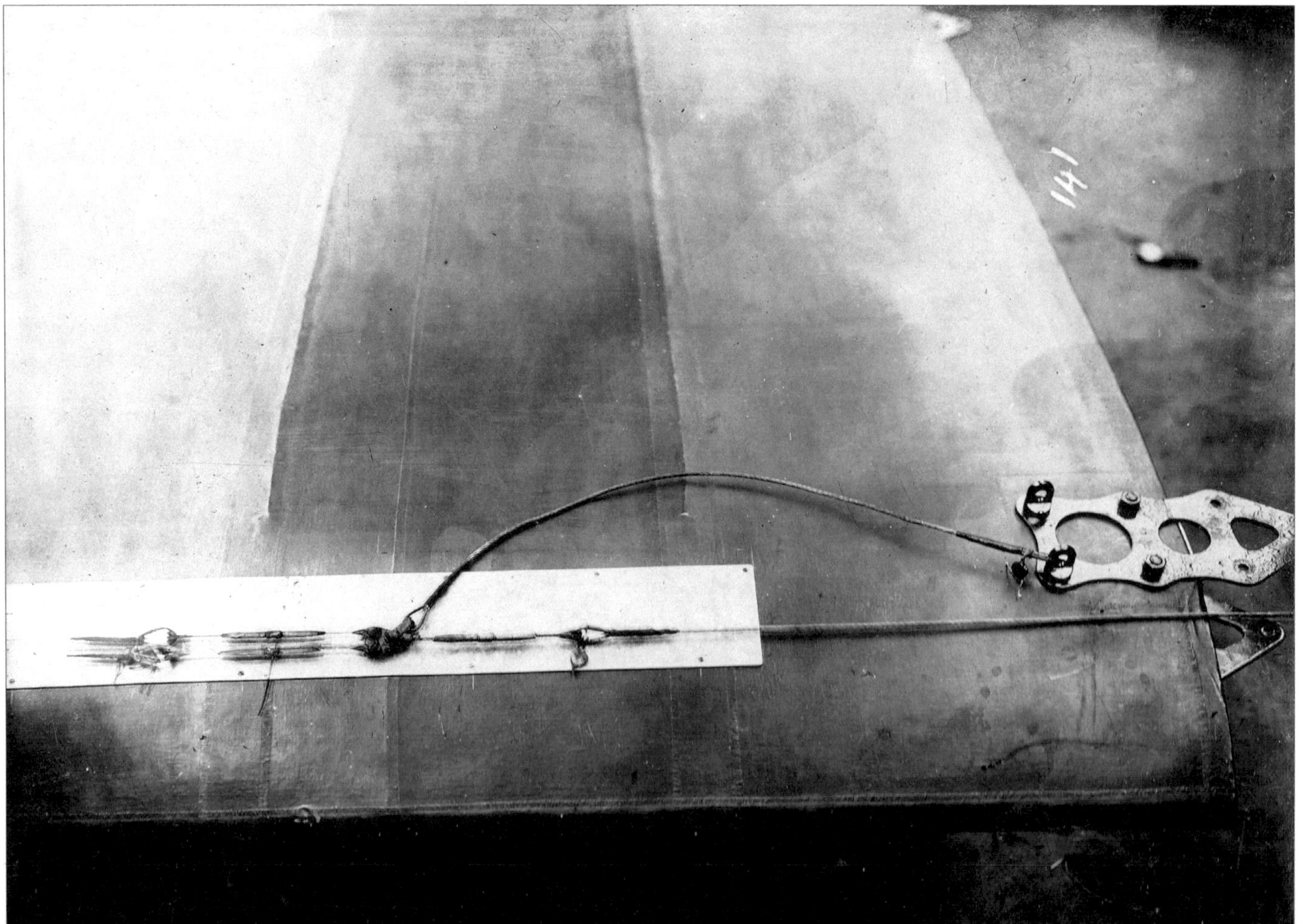

Above: This photograph shows how the bomb rack was fitted to the metal wing rib of a Phoenix built F.3.

Below: F.3 hull with all the ply areas covered, ready for fabric covering.

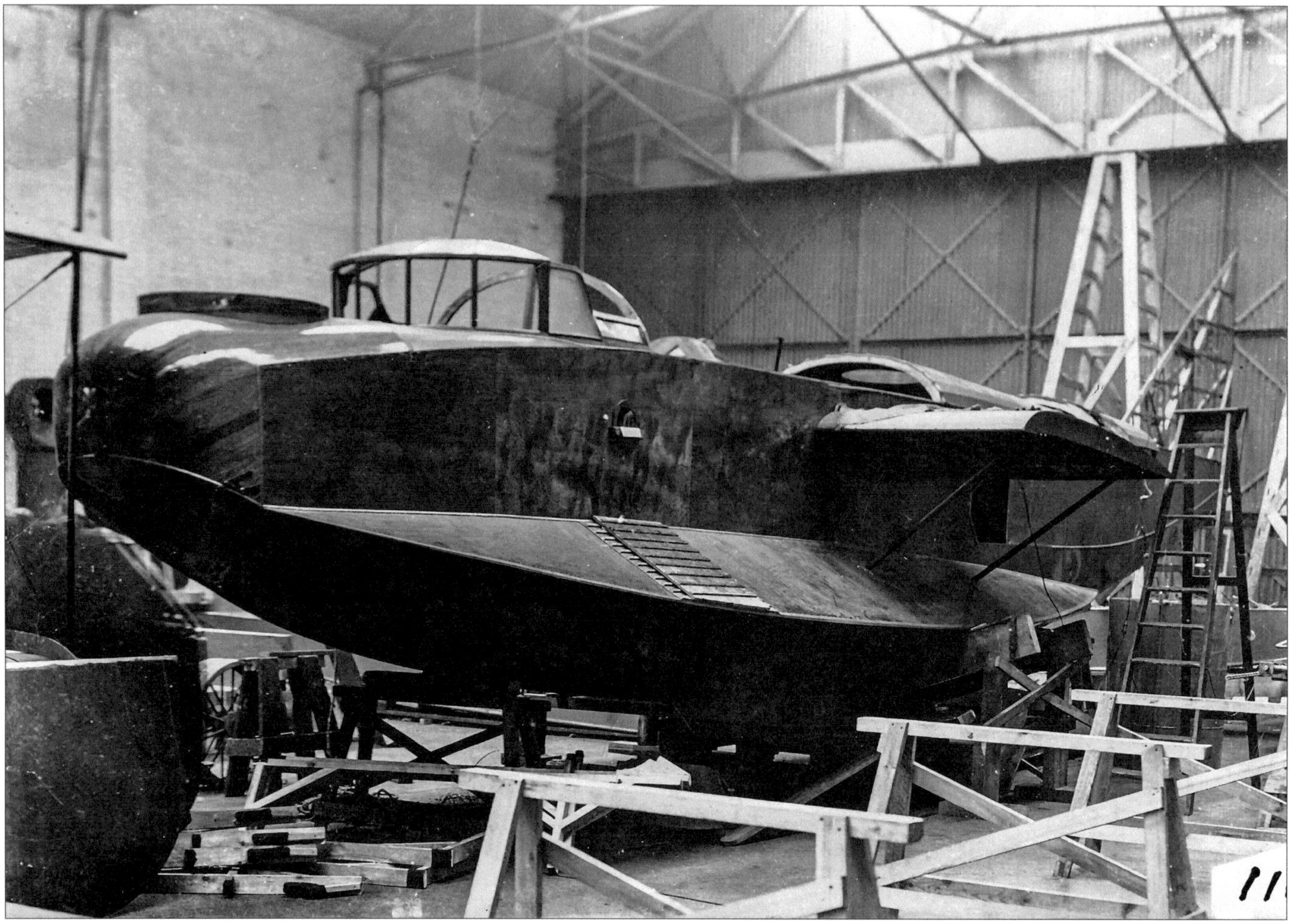

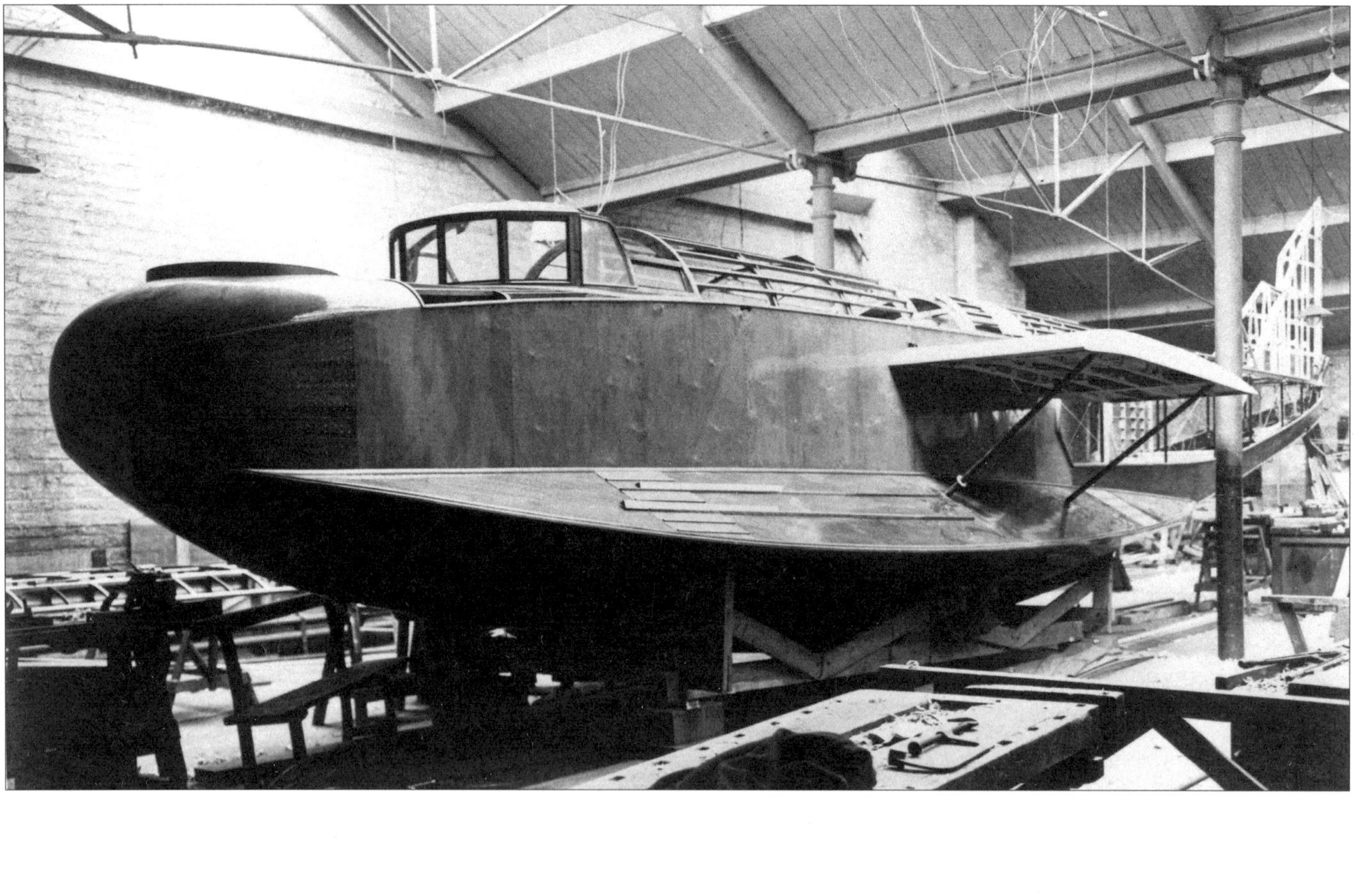

This Page & Facing Page: Completing the F-3 hulls at the Phoenix Dynamo Co works. The hull top and rear of the hull is covered with fabric. The planning bottom and fins were built onto the basic "fuselage" structure.

Above: The elaborate jig for the hull.

Below: This hull is being constructed at Boulton & Paul, November 1917. The basic longerons and spacer fuselage form is well illustrated.

Above: Detail of the rear gunner's section. The hull bottom is not planked while the front section of the hull is.

Below & Top of Following Page: These photograph taken at Boulton & Paul in November 1917 shows the framework for the hull bottom and the completed hull frame has had its bottom planked first. There appears to have been no set method of manufacturing hulls for the F-boats.

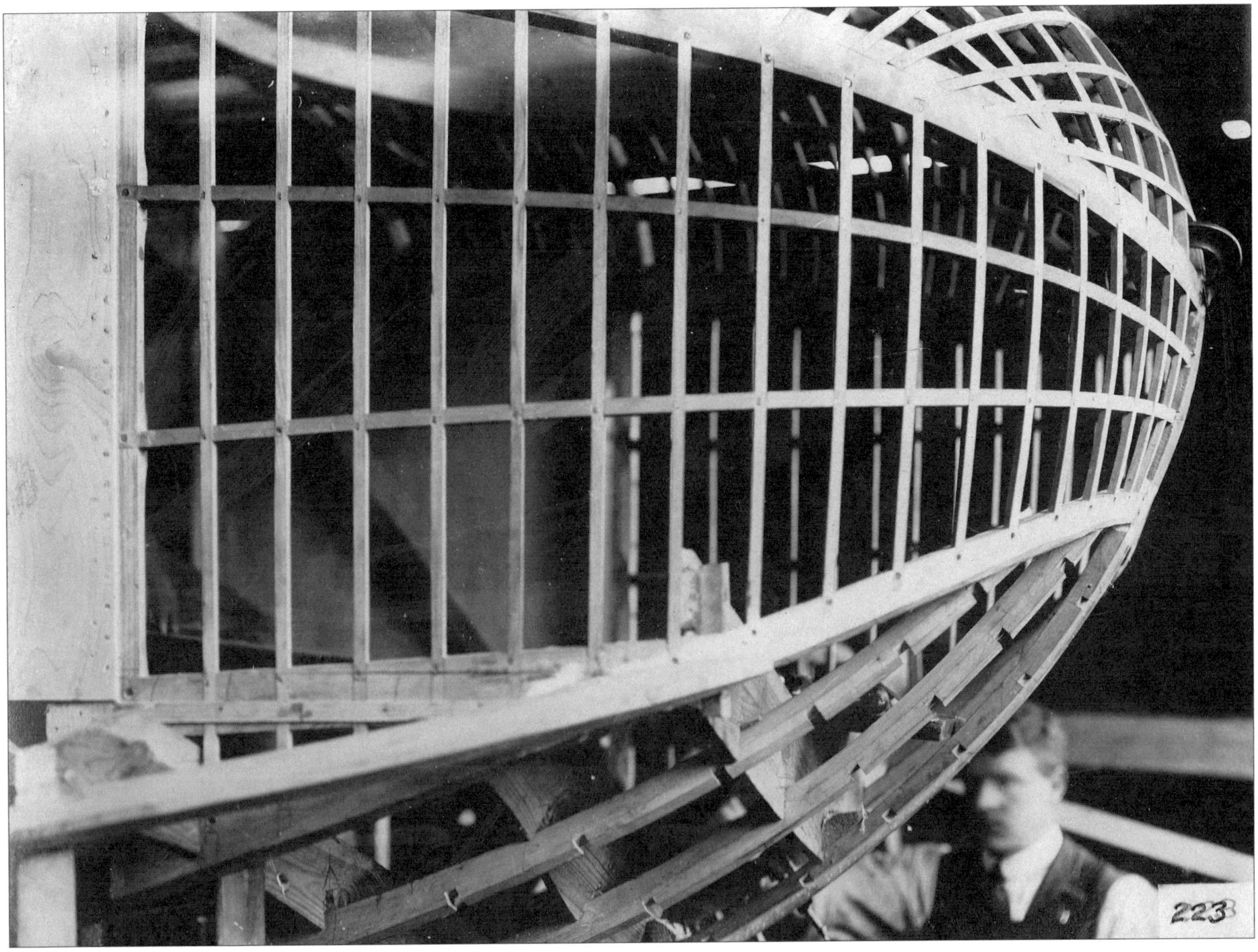

Above: The beautiful wood work at the nose of a Felixstowe F.3 hull.

Bottom of Facing Page: The hull being planked while the hull is upright in the jig.

Below: Damage to the nose of an F.3 that appears to have occurred on the production line.

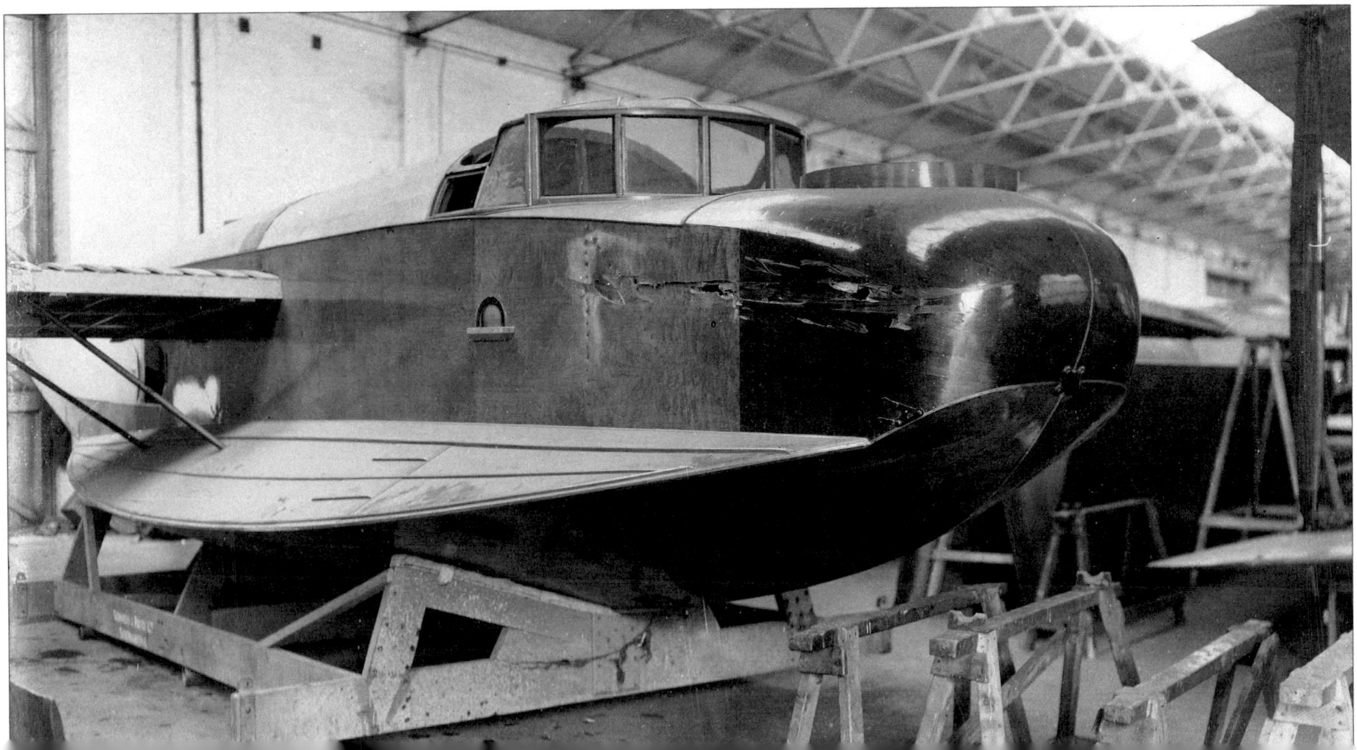

Above: View inside an F.3 hull looking to the rear over the fuel tanks. Note how light the view is due to the translucent doped fabric.

Above: View from the rear gunner's position towards the tail. The fabric is clear doped as the cockades are already painted on the hull sides.

Above: The tail skid or bumper that was fitted to protect the hull from damage when on land.

Below: The late wings with horn balanced ailerons. The upper surfaces were a dark colour with clear doped lower surfaces.

Above: View of a Phoenix built F-3.

Below: Detail of an F.3's wings.

Above: Phoenix built F.3 with the company's construction number, 236, on the hull. Note the style of serial numbering adopted by this company.

Below: Rolls Royce Eagle engines provided the power for the Felixstowe boats. Photographs taken at the Phoenix plant, 29 April 1918.

209

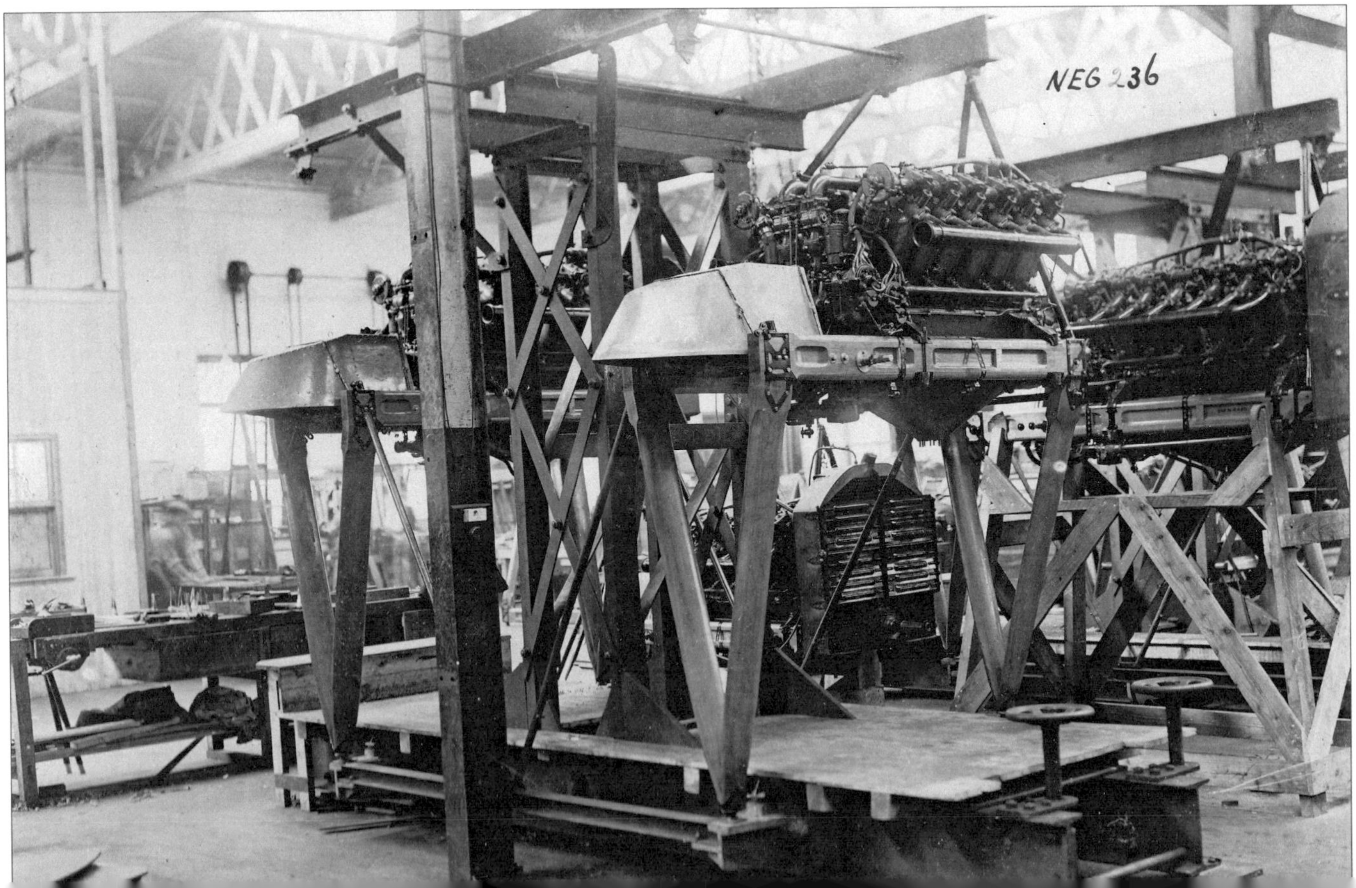
NEG 236

This Page & Facing Page: Rolls Royce Eagle engines provided the power for the Felixstowe boats. Photographs taken at the Phoenix plant, 29 April 1918.

Above: Felixstowe F.3 N4259 with bomb under the lower wing. N4259 was one of a batch of 50 F.3 boats with 375-hp Eagle VIII engines ordered from Phoenix Dynamo Manufacturing Co Ltd (N4230-N4279) and constructed at Preston. Delivered to the Marine Aircraft Depot South Shields in August 1918, it was tested in October. It was still recorded at South Shields at the end of January 1919.

Below: Fully assembled Phoenix built Felixstowe in what appears to have been the standard scheme the company adopted for these boats.

Above: Felixstowe F.3 hull with centre section and engines fitted.

Below Right: Boulton & Paul transporting a Felixstowe hull on its side. The gun ring and cockpit can be seen pressing against the canvas cover. Post-war criticism of the construction program for flying boats noted that orders were placed for flying boat hulls on the Clyde and "orders for the wings at the works over 200 miles away. The hulls were invariably transported to the wings instead of the wings being taken to the hulls. The effect of dragging the hulls of the F.5 and N.4 types over rough roads by night had been that after such transportation the wing-root" spars had become displaced, leading to serious distortion of the whole wing framework. Many minor effects due to the severe treatment which the hulls received on their journey were also evident and gave rise to doubt whether the hull contractor or the wing contractor was responsible." (Capt J.A. O'Brien, RAF. Discussion to D. Nicolson's paper "Design and Construction of Flying Boats", *Transactions of the Institute of Engineers & Shipbuilders of Scotland*, Vol.62, P.308. 29.04.1919.)

Above: The Felixstowe F.2A was built in the USA as the Curtiss H-16 at the Curtiss Aeroplane Co, Buffalo, and at the Naval Aircraft Factory, Philadelphia. This photograph shows H-16 hulls at the Curtiss Austin St Plant's Boat Room. Photo No.821 dated 01.11.18.

Below: A view of the NAF shows less clutter. Each hull has been tagged with the boat's Bureau No.

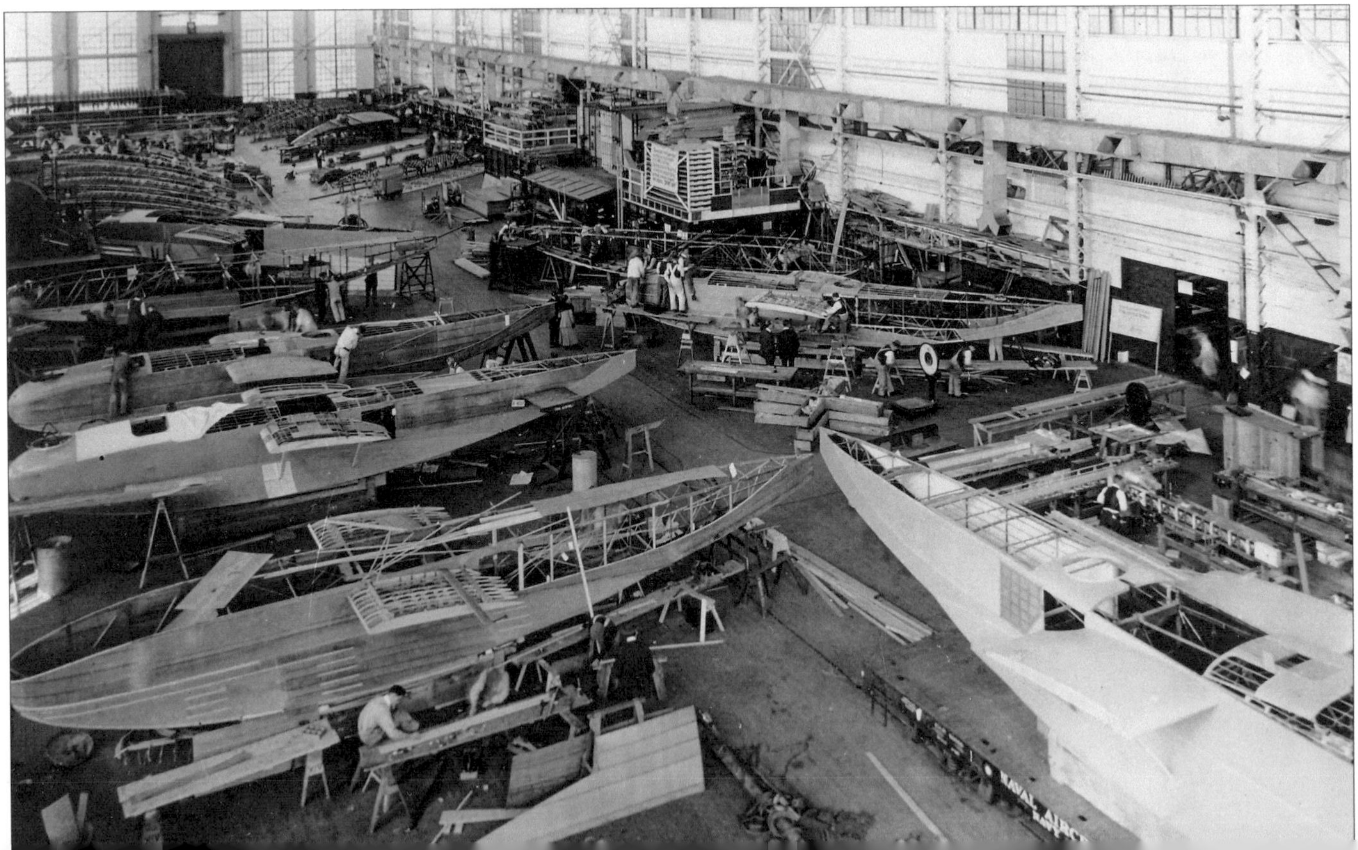

Above: The hull frame is ready for skinning on the left and skinned hulls are coming back down the centre. Bureau No. A-1077 can be read on the visible tag.

Below: Close-up of the nose of an H-16 hull. The change in colour of the hull as it passes through each stage of construction should be noted.

Above: Internal view of the cockpit of an H-16 with dual control.

Below & Facing Page, Top: H-16 boats under construction.

Below: The gap between the cockpit canopy and the rear was a distinguishing feature of the H-16 in flight. The F.2A has a small window that was raised here.

Left & Below: Detail of the nose of an NAF H-16. These boats have a spigot mount for the Lewis gun rather than a gun ring. A Wimperis bomb sight is located on the side of the hull. The hull has been given a gloss varnish finish.

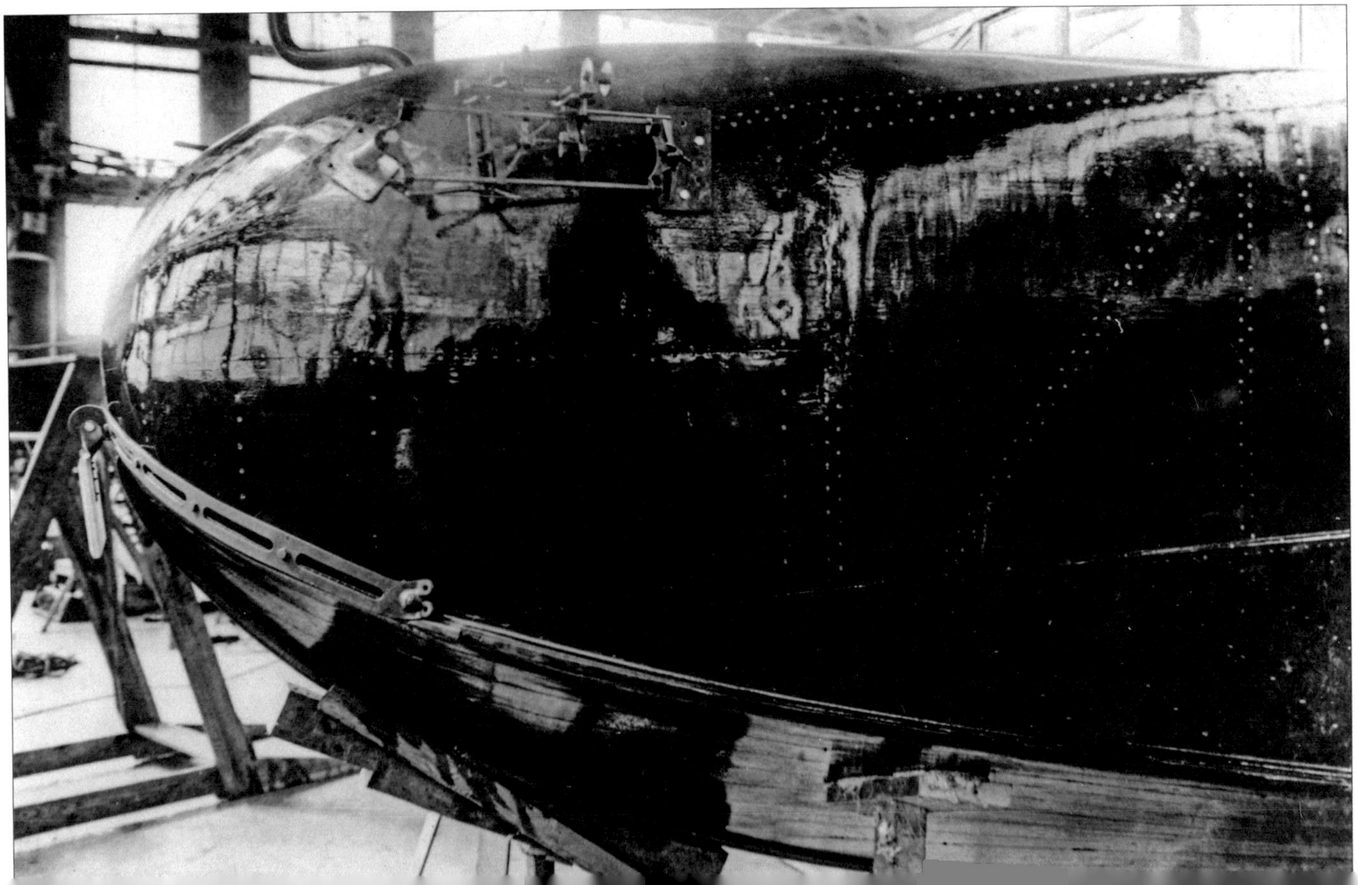

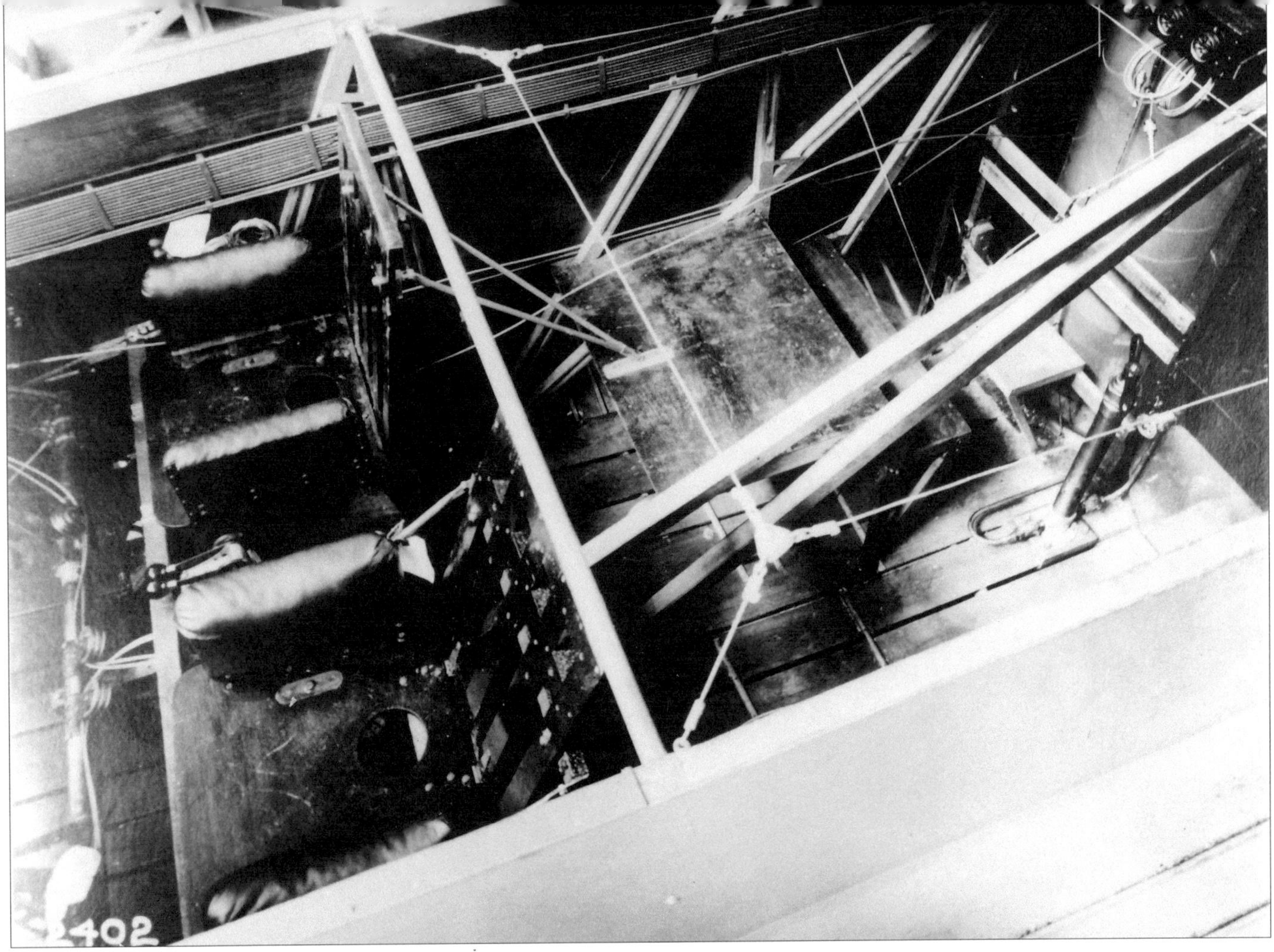

Above & Below: The front cockpit of an H-16 under construction at the NAF. Note the seat frames and dual controls.

Above & Facing Page, Top: The NAF was proud of its employment of women in its construction program.

Right: An H-16 hull that still has to have the rear of the hull fabric applied appears to be in Navy Grey finish.

Facing Page, Bottom: A detailed study of this photograph will reveal much detail of the H-16 hulls.

Above: A Liberty engine packed for shipment.

Right: Internal view showing the Lewis gun stowed on an H-16. Note the flare holder rail above the gun.

Above & Below: The job of packing these large flying boats for shipment was a task in and of itself. The hull was placed in on its side as shown. It is interesting in that this hull has only a single control wheel. Note the "Walking Board" stencil across the hull. It is not known if this was retained when the machines were assembled. The hulls were crated on their side as shown below.

Above: Loading a rail car with H-16 crated parts.

Below: H-16 parts packed and ready to be shipped out. The stencils that can be read on the crate state that they are for H-16 Bureau Nos. A-1044 and A-1077. Curtiss Aeroplane and Motor Corp photo No.1223 dated May 1918.

Above: Crated H-16 on board a ship. The pyramid structure is over the sub-wing as the hulls are crated on their sides. Despite the efforts at making the crates water tight, boats arrived with water damage, but more serious was that the machines arrived with parts missing hence causing delays into moving the flying boats into service.

Below: The bomb sight is on the opposite side of this H-16. The instrument on the nose of this NAF built hull is unknown.

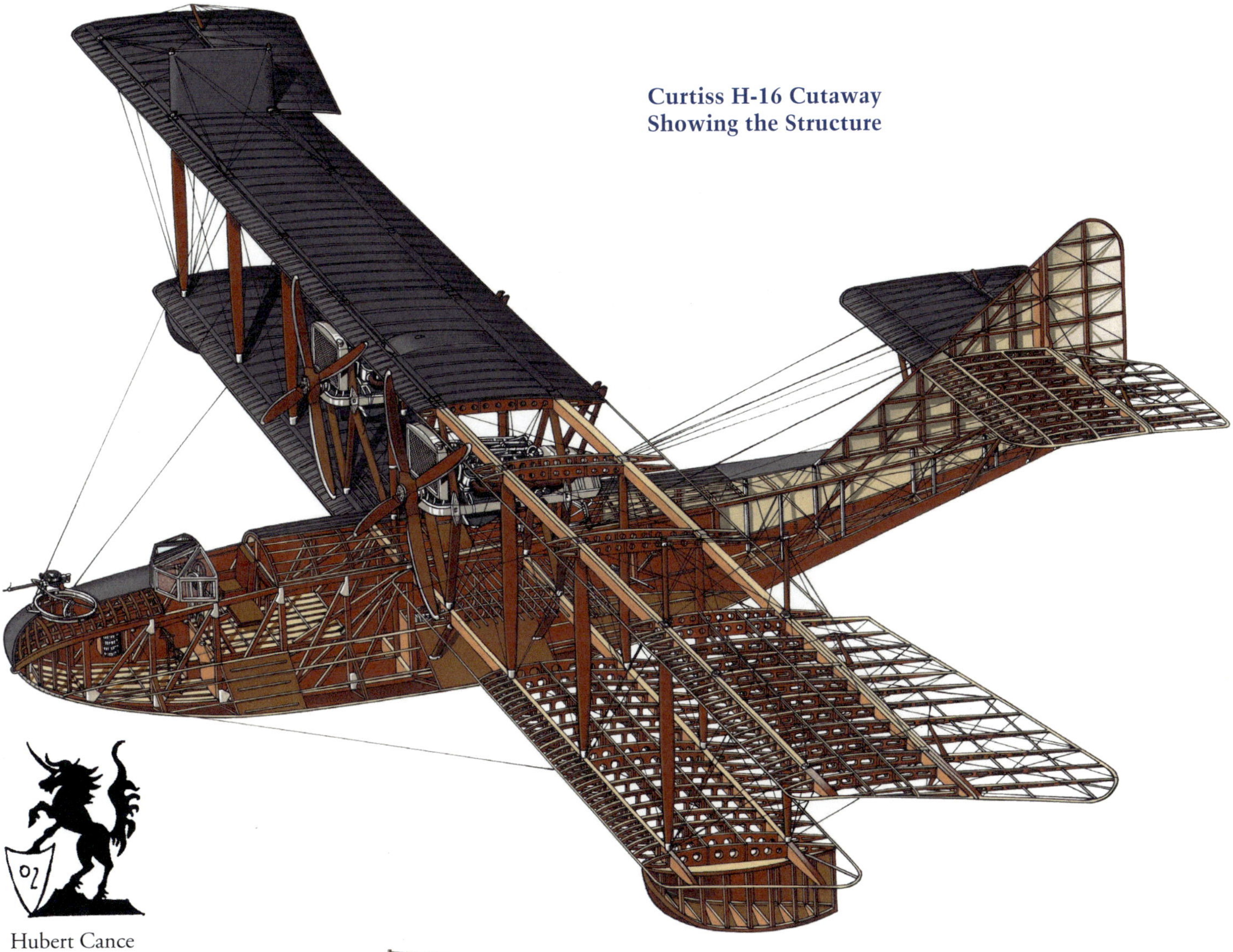

Hubert Cance

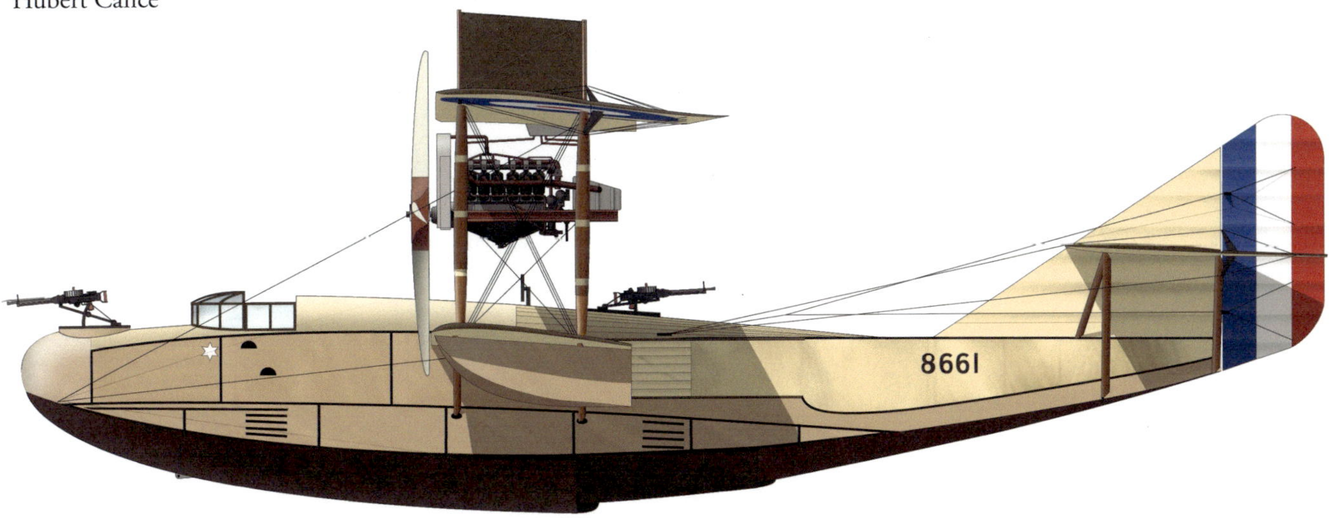

Curtiss H-12 Convert 8661 had a long life, being delivered to Felixstowe on 21 January 1917 and being dismantled as worn out on 20 August 1918. As an H-12 it bombed a submarine and was involved in fights with enemy seaplanes. Converted with an F.2a hull by 6 January 1918, it engaged with five enemy seaplanes in February and shot one down. By June it had been sent to the Felixstowe Experimental Station.

3. The Porte Felixstowe Baby

Above & Below: The first Baby with three pusher engines and short bow. The outboard engines have a diamond shaped structure of struts at the rear for support. Note the three cut-outs in the rear of the upper and lower wings to allow for the propellers. One 300-hp Green and two 310-hp Sunbeam Cossacks. This was the appearance of the Curtiss "three motored America seaplanes which are doing such effective work in submarine hunting for Great Britain," in photographs published in magazines and journals of the time.

Above: Hull of Baby 9800. Note the footsteps let into the hull just forward of the wing; the wing root attachment around the spars that pass through the hull. There is a cut out in the upper decking of the hull behind the cockpit. This is shown on drawing T.902 of 9800's hull, but does not appear on production Babys.

The Porte Baby, a bigger machine than any built and flown in this country until 1918, and this boat was produced in 1915 and flown in 1916. Although it did little useful active service work, it set other designers to thinking, and was the father and mother of all big British aeroplanes and seaplanes. When fully loaded it weighed about eight and a half tons, but no scales big enough to weigh it were obtainable in the service.[1]

So Sqdn Ldr Thedore D. Hallam or *Pix,* described the Porte Baby. According to Commodore Murray F. Sueter, RNAS, the "building of this large experimental flying boat...was carried out during the same period as the three series of experimental modifications on the Curtiss machines, and the experience obtained thereby was both incorporated and confirmed by the "Porte" flying boat."[2] The experiments referred to were made by Porte at Felixstowe with various types of hulls fitted with Curtiss wings and tail units. While the experience of these hulls was incorporated in the Baby's hull, "it was not possible to take advantage of the results from the Porte I."[3] This latter hull was to be developed into the successful Felixstowe series of flying boats. At the time Porte was carrying out his experiments he was also building the Baby and it was completed before the results of the F.1 experiments were known. (See Chapter 2)

The origin of the Baby is not as clear cut as this would suggest as photographs appeared in the journal *Flight* for 21 September 1916, and *Janes All the World's Aircraft* for 1917, depict what appears to be the Porte machine but it is identified as a Curtiss "Super-America" flying boat that was "capable of rising from and alighting on, very rough seas". The boat was equipped with three "high-powered Curtiss motors" that were installed in pusher configuration.[4] No mention of this boat as a Curtiss product is found in Bowers' *Curtiss Aircraft 1907-1947,* however a drawing of a three engined pusher flying boat labelled the Curtiss three-engine Cruiser was published in a contemporary book.[5] So many features of this "Curtiss" boat are common to the Baby that it is probably the Baby in its original form. Notwithstanding the origin of this flying boat, it was an important step in the development of the large flying boat as a weapon of war.[6]

According to Major W.F. Vernon "H.M.S. Baby" was designed in 1914 and launched in 1915, while the Felixstowe Daily Reports for 19 November 1915, records that the "triple-engined "America" was launched but no flight attempted owing to trouble with Sunbeam motors."[7] The flying boat that emerged was the largest built to that time in the UK, and perhaps the world, so it was automatically dubbed the Porte "Baby" although referred to as the Porte Flying Boat in the official publication "Types of British Seaplanes." The serial number 9800 was allotted to the prototype. The first flight was on 20 November when it "got off the water easily," although the Sunbeam engines were running badly. The following day the "Big Boat" "gave greatest satisfaction." On 17 December another flight was made but again engine trouble prevented tests from being fully carried out.[8]

The hull was experimental as noted above, and was the result of experiments that Porte had carried out with full sized hulls in order to develop a flying boat that could operate over the North Sea. The hull was generally similar to that of the *Small America* boats and was 14 feet wide and 53 feet long.[9] It was built out of cedar and mahogany planking with pitched elm timbers. Although rather heavy, the hull was a fine piece of boat building.[10] The fins decreased in thickness and ended up

Above & Below: More views of the hull of Baby 9800 showing additional details. Above shows the Davis gun mount.

Above: Solid compression rib of Baby wing.

almost horizontal at their trailing edge. They projected from the side of the hull 2½ feet, and although supported by a strut, they were a source of weakness and liable to get broken when getting off or alighting in a moderate sea. The planning bottom was slightly concave increasing in flare forwards. At the step the angle was 7° to the horizontal but it got steeper towards the bows. The tail was flat underneath at the sternpost but became slightly V shaped at the step.[11]

The Baby was a two bay biplane with an enclosed cockpit. The unequal span wings had no stagger, and the upper wing had large extensions that were braced by kingposts above the outer pair of interplane struts. Large ailerons that increased in cord towards the wing tips were mounted on the upper wing only. The lower wing had rounded wing tips and floats were mounted under the outer interplane struts. The three engines were supported on a system of V struts, the inter-wing two were mounted as tractors, and that on the centre line, a pusher.

A large triangular fin was mounted on the deck of the hull. The rudder had a rounded trailing edge, the bottom being in line with the bottom of the hull. A two piece elevator was mounted. The horizontal tailplane was supported by two struts on each side.

> *On rough days a great deal of spray was thrown up when accelerating to get off. Consequently it was decided to lengthen the hull by 3 feet in front of the wings. This was a great improvement.*
>
> *When finally put into service this machine behaved well; her performance on the water being far ahead of anything which had yet been produced.*[12]

On account of the large unbalanced control surfaces and the great size and weight of the machine it was very heavy on the controls. Sperry electric servo motors were fitted to supplement the pilot's strength. They were very ingenious but needed much skilled work to keep them in order. It was possible to throw them out of the control loop and fly the boat by hand. This was not as serious drawback as it sounds as the pilots could easily relieve each other.

A number of armament tests were tried including a Davis gun, Hotchkiss guns and torpedoes being fitted at various times. Porte reported on 28 March 1916, that "Two 14" torpedoes are now slung on the machine. This test will be made as soon as weather permits."[13] The most remarkable experiment was the loading of a submarine mine that was to be launched through a port in the side of the hull. This experiment was never completed.[14]

In the 28 March report referred to above, Porte noted that the bad weather had greatly hampered the testing of the Baby. Results achieved to date were listed as follows:

Speed	60 knots. 50 knots with wing motors only.
Climb	Uncertain, about 200 ft/min. Only reached 2,200' due motor trouble.
Useful Load	4,500 lbs. When this test was made the engines were only developing about 650 HP instead 800 HP.
Servo Motor	This was underpowered and caused considerable trouble. "We are now fitting a larger type and a different system. Controls have been operated by hand in the meantime."

A duration test was carried out on 19 May 1916, with the "object of ascertaining definitely the range, etc., of this machine." Porte recorded that the seaplane left the water at 11:19AM and alighted at 7:25PM, a total time of 8 hours 6

Above & Below: Uncovered Baby lower wing. The solid compression ribs are well shown. (This selection of photographs accompanied M. Sueter's report on the experimental work done by Porte at Felixstowe.)

Above & Below: Baby 9800 with short bow on the Felixstowe slipway. The cockade is marked well out on the upper wing (starboard only) and did not overlap onto the aileron. No cockade has been applied to the hull, nor have rudders stripes been added yet. The machine has had the engine layout changed to two tractors with a central pusher. The cut-outs in the upper wing have not been filled in at this time. Note the camouflaged hangars in the background. (Nicolas Cooper via SEAWINGS)

Above: Judging from the colour scheme and lack of national markings this Baby is 9800. The cut-outs in the wings have been filled in.

minutes in the air. The flying boat landed with 57½ gallons of petrol remaining out of 438½ gallons. It was estimated that this would have allowed, with the 20 gallons wasted before taking off, for 1¾ more hours flight. The "actual distance made good" was about 421 nautical miles. This was an average air speed of about 55 knots.

The load comprised the crew: Cdr Porte; Flt Cdr Hope Vere,

FSL Scott, CPO Tadman and L.M. Coates ...	830 lbs
496 gals Fuel	3,621 lbs
Lewis gun & Ammunition	342 lbs
Tools & spares	54 lbs
Food, kit, etc.	125 lbs
Total Load carried	5,038 lbs

There was 20 gallons of oil remaining and allowing 160 lbs for the extra "passenger", and the unused fuel it was calculated that this was about 900 lbs which could be made up in bombs or a 6 pdr David Gun and ammunition. "This puts this machine down at 8 hours fuel and 6 pdr David Gun with 30 rounds or 900 lbs in bombs. If you allow 400 lb for each hour's fuel the armament can be increased proportionally."

Porte continues:

The behaviour of the machine in the air is excellent, being quite stable. It is possible to fly her for considerable periods without touching the controls.

The Servo-motor broke down at five hours, necessitating using controls by hand; this proved all right, but would be very tiring on a rough day. The fault was only a small one due to bad design, but this has now been remedied.

The motors ran beautifully, although after five hours the Port Motor popped considerably, this passed away at seven hours. The Starboard Motor ran perfectly throughout. The latter was fitted with KLG spark plugs, the other two were Lodge of two kinds. There was no comparison between the plugs when examined, proving conclusively that KLG are by far the best for long flights.

The motors were opened out to full power at the conclusion of the flight and they all ran perfectly at full revolutions.

For some reason unexplained as yet the Port Engine used 2 ½ gals of oil or ½ gal per hour as compared with 5 ½ gals by the other motors. Yet this motor has suffered no damage and ran well. This points to the fact of those other motors being over-oiled as a general rule.

The cooling was perfectly satisfactory, the overflow pipes from radiators are led to a cup in the hull, so that if the water boils or overheats it can be seen immediately. This would seem to be a good fitting to place on all machines.

The petrol consumption is somewhat high on these motors, due to the fact of a comparatively low compression. If this could be increased satisfactorily the consumption should improve.

In these big machines the weight of the engine is not so important as its efficiency and reliability. All our experience up to date goes to prove that these two qualities have not gone hand in hand. Low efficiency motors have been by far the most reliable.

Above: 9800 at a later stage with extended bow, dark hull, and large cockades that overlap the ailerons, rudder stripes and serial on fin. There is a cut-out in the upper wing centre section. The large aileron control horns stand out in this view.

The Rolls-Royce combines greater reliability with moderate efficiency. If it was possible for this firm to increase the efficiency of the motor at the expense of slight increase of weight, then this motor would certainly be a wonderful motor. As far as my experience at present goes, I would say that for reliability it is a motor apart.

The above remarks refer purely to a long duration motor: for short duration the weight of the motor is all important and its efficiency not so important.[15]

It was reported that two 14 inch torpedoes had been carried, one on each side under the wings. This was on No. 9800, the prototype. There is no record of them being launched. Maximum speed was 68 knots, and 50 knots with the wing motors only and a light load. Climb was 6,000 feet in one hour but this was expected to be improved with a different arrangement of carburettors. Useful load was 5,000 lbs.

In his report and presentation to the Advisory Committee for Aeronautics on the experiments carried out at RNAS Felixstowe, Commodore Sueter included a copy of Wing Commander Porte's report on the Baby. In this report Porte noted that it had flown for eight hours at an average speed of 55 knots with 1 hours fuel remaining in the tanks. Load had included "5 passengers and Lewis Gun etc., but no bombs." A normal crew was two pilots and two others. It was usual to include under "passengers" any crew other than the pilots. However it may well be that he is referring to the same duration flight of 19 May as detailed above. Porte continues to detail the history of modifications to the boat:

The machine was originally designed for 3 motors of 300 H.P. each – making 900 H.P., the estimated weight of each motor being 850 lbs. without water and radiator, or about 1,050 lbs. in all, making 3,150 lbs. for the three motors.

As it eventually transpired three engines of 240 H.P. of about the same weight were fitted, making a total of 720 H.P. instead of 900 H.P.

The hull was fitted with a single step in a vertical line with the centre of gravity of the machine. The bottom in front of the step had a slight "V" in cross section, this "V" being somewhat hollow towards the keel, otherwise straight. The bottom of the tail portion was also a "V" at the step, flattening out to nothing at about 8 feet from the stern post. The usual fins were fitted on the sides of the forepart of the Boat. The overall width of the bottom at the step was 14 feet, the width of the main portion of the hull itself being 7 feet.

Trials were commenced with two Rolls Royce engines fitted as tractors on the wings and a 280-hp Green engine fitted over the hull as a pusher. The wing motors rotated inwards.[16]

When the Boat was first taken out it was found that she planed very easily but was difficult to make rise from the water, due to drag on the tail, the tail touching every time we tried to get

Above: Front view of 9810 at Felixstowe, 11 Jan 1918. Note large cockade only to starboard wing.

off. This I circumvented by jumping her off, which was made possible by the 20 mile an hour breeze which was blowing at the time. However, it was evident that there was a mistake somewhere.

The machine was also considerably tail heavy, due to the fact that the centre of gravity proved to be much further aft than was anticipated.

When the machine was brought ashore and hangared, Porte had the fore and aft angle measured very carefully and it was found to be 5¼° instead of the designed 7½°. This was found to be due to mistakes on the drawings. "We then took a slice off the underpart of the tail, that is from the step to the sternpost, leaving the same depth of step, making the tail angle 7½° with the front of the boat. This was accomplished without difficulty."

While these modifications were being carried out, the opportunity was taken to move the centre of gravity forward. The two tractor Rolls Royce engines were shifted forward so that their centres were over the front spar and this naturally placed the airscrews some way in front of the wings instead of in line with the leading edge of the wing as formerly. The diamond-shaped structure of four struts at the rear of the engines was replaced by a normal interplane strut between the upper and lower rear spars in line with the engine. Trials were then continued.

It was found that the boat rose easily from the water without the tail catching as previously. On the other hand, she did not plane so quickly due to the fact that the angle of attack on the front of the boat had been increased. The two air pipes that had been fitted to break up the suction behind the step were found to be unnecessary and were removed. The machine was now in balance and although the tail carried a certain amount of weight, the machine was stable fore and aft.

It became apparent that the boat tended to wallow in a following sea, and the bow was lengthened three feet. This modification improved the boat's "performance on the water enormously."

A new type of wing tip float was tried. This float's bottom was double concave in section; the centreline of the keel was very low and pointed. "The result was that, even in a seaway, practically no shock was felt at all when the water struck the floats." When lying on the water head to wing, the machine was stable and remained on an even keel with the wing tip floats clear of the water.

The machine was fitted with conventional and servo-motor controls. The servo-motors gave trouble and thus hand control was used fairly frequently. "This meant considerable exertion on the part of the Pilot and prevented the machine being flown in bad weather for any length of time. After a time the servo-motors became fairly reliable and there was "no reason why this instrument should not become absolutely so."

The Green engine gave a good deal of trouble mostly due to the "very crude oiling system." Eventually, the Green was replaced by a Rolls Royce engine and "from that day forward

Above: 9800 with Davis gun mounted. Note how the rear cut-outs in the upper wing at the two outer engines have been filled in. The tonal values of the hull in this view are noteworthy. The cockades on the wings are well outboard and do not overlap the ailerons.

we had practically no engine trouble."

A very large vertical fin was required to compensate for the large amount of frontal side surface of the hull. One drawback of this was that the slipstream of the central engine acted obliquely on the fin and had a tendency to turn the machine. In order to eliminate this, two small rudders were fitted at first and while they were successful, Porte did not consider this a satisfactory arrangement and the fin was hinged and was controlled from the cockpit. "This proved successful."

The distance from the fuel tanks to the gravity tanks above the motors was about 15 feet and at first great difficulty was experienced in pumping fuel to these gravity tanks. "This was eventually overcome by putting the fan driven petrol pumps outside the hull at a level which brought these to within two feet of the bottom of the main tanks."

With the Green engine in the central position the speed was 60 knots, and on substitution of the Rolls Royce engine for the Green, and improving the streamlining generally of all mountings, radiators, etc., the speed was increased to 68 knots, without streamline bracing wires. Minimum speed was 47 knots with full load.[17]

No. 9800 the prototype Baby, was subject of an unusual experiment wherein a single-seat Bristol Scout C biplane, serial 3028 from HMS *Vindex*, was mounted on the top wing of the prototype Baby and air launched. This combination was undertaken to provide the fighter with a means of encountering the Zeppelins that were raiding England in 1916. The Porte flying boat with its long endurance would carry the scout on patrol until an airship was sighted whereupon the little fighter, that had an endurance of only two hours, would be launched to pursue and hopefully destroy the airship.

The wheels of the Scout rested in two steel channels supported by the centre engine struts. The rear of the fuselage was held in place by a hinged steel frame supported by the back spar of the seaplane's upper wing. "A "Rubery-Owen" slip was attached to the axle *(of the Scout)* and worked from the pilot's seat." A successful launch was made on 6 April 1916.

Once the seaplane was in flight the engine of the Scout was run at full speed and the back frame pulled down and out of the way, the machine being supported only by the front wheels, and under control by the elevators. "When the pilot of the Scout *(Flt Sub-Lt Day[18])* was ready, he set his machine at five degrees inclination, and released the slip. The Scout immediately jumped up about 10 feet clear, and made off."[19]

Above: 9800. The cut-out in the centre of the upper wing and small cockade are apparent in this view. (via H. Alderson)

According to an eye-witness it took off "like a dove from a roof."[20]

It had been recorded that only one air launch was carried out but another was recorded for 17 May 1916, with Sqn Commander Porte piloting the flying boat with Flt Lt Hope and two crew, and F/Lt M.J. Day in the Bristol Scout. This flight from Felixstowe was also successful, if indeed it was a second test, with the Scout being released at 1,000 feet over Harwich, from where it flew to a landing at Martlesham Heath. In the notes he penned to Maj Vernon's Report Porte noted that a Sopwith Pup was launched. "This effort was most successful, the Scout releasing herself without any difficulty and flew back to land." It is possible that Porte's recollections are incorrect as no other reference to a Pup being launched has been found to date. Porte considered that it "would seem a great pity in the light of other experience that the scheme was never used in active service."[21]

In a Report dated 27 July 1916, Porte gave the maximum speed as 68 knots and with a half load the machine could climb 6,000 feet in an hour. "These performances are not extremes and all the machines should be able to accomplish them."[22]

A report on No. 9800 dated 25 November 1916, states that the seaplane had been reconstructed and fitted with 310-hp Sunbeam engines as left hand tractors and one 250-hp Rolls-Royce as a left hand pusher. This reconstruction is presumed to be that referred to above. On the first trial of the modified aircraft it was found that owing to the three engines revolving in the same direction "a blast was forced onto the tail fin and rudder which gave the machine a tendency to turn to the left. This tendency was so violent that it was practically impossible to overcome it by means of the rudder." The central engine was replaced by a right hand pusher and with this change the machine was balanced directionally and was "now considered satisfactory."[23]

Twenty Porte Babies[24] were ordered on 16 February 1916, under contract C.P.104214/16 (serials 9801 - 9820) with the hulls being built by May, Harding and May of Southampton Water. The machines were assembled at Felixstowe and Killingholme. The engine mountings were the same as the modified prototype. At least ten of the flying boats and all of the hulls were completed.[25] This number is confirmed by a note in a file describing the aircraft and seaplanes in use by the RNAS wherein it is noted that the Porte Baby had

> *not proved as handy as the "Large America" type and no more are being ordered after the 10 on order are completed.*[26]

While Rolls Royce engines were to be the standard installation, at least one of the production machines (9801) had a big water-cooled 260-hp Green V-12 as the central engine. With the Rolls Royce engines a total of up to 1,050 hp was available and the loading reduced to where a good performance was possible. One tested at the Isle of Grain gave a top speed of 76 knots.

Above: 9802 under erection. (This photograph accompanied M. Sueter's report on the experimental work done by Porte at Felixstowe.) 9802 was delivered to Felixstowe for erection on 2 August 1916. By October 1917 it had two 320-hp Sunbeam tractors and a 250-hp Rolls Royce as the pusher engine. The machine was stored at Felixstowe around December and deleted in July 1918.

A letter in February 1917 discussed the policy of the RNAS and what types of operations could be carried out by the service noted that the RNAS required a Wing of long distance bombers to operate against blast furnaces supplying steel for submarine manufacture, while considering that operations with the BEF supporting the Army were "a purely military operation in no way connected with the Navy." The commissioning of the Seaplane Carrier "Argus" was a high priority however notes indicate that it would not be ready by November 1917 as stated. Under the discussion of Large Seaplanes only two were mentioned, the *Large America* and the Porte Flying Boats. The Porte boats were described as

> *large boats of 750 H.P.*
> *Radius of approximately 400 miles. Armament: Machine Gun with either a 6 pounder Davis Gun, or can be fitted with 2-500 lb. bombs. Speed: 60 knots.*
> *Number in Commission: 2.*
> *Number erecting at Felixstowe and Killingholme 7 at approximate rate of one per week.*
> *Remaining on order: 11. Rate of delivery: one per week.*
> *These Large Boats require large sheds and heavy slipways and, at present, can only be accommodated at Killingholme and Felixstowe, and it is proposed to station ten at each place for extended patrol work. No more of these machines are being built, being superseded by large America Type.*[27]

Although regular patrols were not attempted, they were flown operationally over the North Sea on the Spider Web patrols from Felixstowe and Killingholme from November 1916. E.M. Ackery was a Flt Sub-Lt in December 1916 when he was sent to Killingholme to learn to fly seaplanes. He remembered that the big Porte boats were used for patrolling the North Sea but he never saw one take off in his time at the station.[28]

On 1 October 1917, Baby No. 9810, piloted by Flt Cmdr N. Shoto Douglas with Flt Lt Basil D. Hobbs and four other crew members, was attacked by three German aircraft, two seaplanes and a landplane, near the North Hinder Light Vessel off the Dutch coast shortly after 4 pm.[29] T.D. Hallam recorded the incident in his wonderful book, *The Spider Web.*

> *Well on in 1917 sundry young pilots took the Porte Baby out for a joy - ride, and presently found themselves off the Belgian coast being attacked by a Hun land-machine and two fighter seaplanes. Two out of the three engines were shot about and the big boat had to come down on the water. The Huns circled around firing at it until their ammunition was exhausted, and then returned joyously to Zeebrugge to report the total destruction of a giant flying-boat.*
>
> *But while the tracer bullets were playing about, the crew were lying down in the bottom of the boat watching the splinters fly.*

Above: Baby 9810 in a hangar at Felixstowe. Note the wing of another Baby over the tailplane. The boats were moved into the hangar sideways in order to shelter as many as possible. 9810 was at Airco by February 1917. It is assumed that the flight organs were manufactured by Airco and the hulls by May, Harding and May. By October the machine had three Eagle VIII engines when it was forced to alight in the North Sea by hostile aircraft. Returned to service this boat survived until December 1918 when it was apparently stationed at the Isle of Grain..

When the Huns departed the crew repaired the engines, started them up, and all night long taxied on the water across the North Sea. The much - chastened pilots beached the boat, in the small hours of the morning, on the coast of England, near Orfordness. A sentry, believing, as he explained later, that at last an invasion of England by Zeppelin was being attempted, fired on them, but was eventually pacified. The crew arrived at the station very tired, very black, one of their number with a bullet hole in him, but cheerful.[30]

Hobbs had evaded the attacking enemy aircraft for about 20 minutes until the port and central engines suffered enemy fire and stopped, causing him to alight on the sea. Spikings had continued to keep up his fire despite being scalded when the engine near him had been set on fire. Once on the water the enemy machine gunned the boat wounding Davies, the W/T operator. Despite his injuries Spikings managed to work on the engines for several hours such that they were able to taxi towards the English coast. They managed to reach Sizewell Gap on the Suffolk coast from where they were towed back to Felixstowe.[31] As a result of this attack the Babies were never again used where there was the risk of interception by enemy aircraft.[32]

This fight was the combat debut of the Brandenburg W.12. A sortie by three Friedrichshafen FF 33L, an Albatros W.4 and Brandenburg W.12 *Marine Nummer* 1183, left Zeebrugge at 1600 (German time). After one FF33L dropped out and alighted with engine trouble, another returned to summons help. The three remaining aircraft encountered the flying boat and attacked. The British crew probably misidentified the W.4 as a landplane due to its small size compared with the other two aircraft. The Germans claimed the British flying boat was shot down with bullets in its left engine and radiator. The W.12 was credited with the destruction of the flying boat, the pilot, *Oberleutnant zur See* Friedrich Christiansen, claiming his second victory.[33] This was to be the start of the increasing activities of the Zeebrugge floatplanes and their Brandenburg biplanes and monoplanes against the British patrolling flying boats that lasted until the end of the war.

Arnold D. Massey recorded that he had flown a Porte boat from Killingholme to Catfirth, about seven miles north of Lerwick. The hangar wasn't ready and an 80 mph gale broke the upper wing extensions. "The plane never made a patrol up there as the replacement parts were never received."[34] This was 9807 that is recorded at Catfirth early in July 1917 and was deleted the following December.

The servo motors were a constant source of trouble and even when they worked, the controls answered so slowly they were most tiring to fly in any but the calmest of weather. The fin corners broke on a number of occasions until they were strengthened.

Above & Below: The original caption stated that this Baby was at Scarpa Flow in 1917. Unfortunately the serial cannot be made out on the original photograph.

Above & Below: 9807 after suffering major damage to the wing extensions in a gale at Catfirth in the Shetland Islands on 14 June 1918.

Above & Below: 9810 with gunner's cockpit in the bow. The upper surfaces of wings, tailplane and hull are a dark colour. Note the neat light coloured (white) demarcation line along the edge of the fins to separate the top and bottom of the hull. It is assumed that the bottoms of the hulls would be painted with an anti-fouling paint. These photographs were taken at Dundee while en-route to Houton Bay. (via M. Tuckey)

This Page & Facing Page: This Killingholme Baby also has the nose gunner's position, however the serial cannot be read on the original photographs. Note the 520-lb bomb being installed under the port wing and the fuel pump propeller. (via M. Tuckey)

Above: Baby 9805 has its cradle well marked. 9805 was sent to Killingholme for erection in December 1916. Forced to alight while on a flight to Dundee it was towed to Newton Bewith in May 1918 with a damaged starboard wing and float. It was deleted the following July.

Above: Baby moored out to a buoy. (via P. London)

Above: Baby 9807 on its way to Scapa Flow, May 1918. Delivered to Killingholme for erection it was accepted for service in March 1917. Re-engined with Rolls Royce engines it served at Catfirth in July 1918 and was deleted in the week ending 19 September.

Below: Unidentified Baby on slipway.

Above & Below: The serial of this Baby cannot be determined. Note floatplanes in background. (via M. Tuckey)

Above & Below: 9800 with Bristol Scout 3028 mounted over the top wing. Porte thought that this method of attacking Zeppelins had great promise and it was "a great pity in the light of after experience that this scheme was not used in actual service."

Above: Cockpit of Porte Baby 9810 (Source: Secret Technical Information , 15 May 1918).

One of these Babys went to the Firth of Forth where she was out of action for several months due to engine trouble until the hull became waterlogged. Another got as far as the Orkneys where she broke free of her moorings and was blown ashore but without damage. She was refloated and continued to carry out patrols.[35]

"It was decided not to proceed further with the development of the type, as the water performance was much inferior to the Porte II, and the type of hull construction (was) weak."[36] The machine was slow and underpowered. It was hard to defend having a multitude of blind spots. While at least one had a Lewis gun on a gun ring in the forward cockpit, the majority were armed with Lewis guns firing through windows and ports. The prototype, No. 9800, had a Davis Gun fitted at one point but no record of its use has been discovered to date. The performance of the Baby would not have let it be an effective anti-Zeppelin weapon. Capt David Nicolson considered that although the Porte Baby "was constructed under the best conditions with respect to material and workmanship, they were inherently weak in the bottom, especially at the step, owing to the faulty design of the keel, which was of spruce, and very small in section at the tail."[37] Porte considered that the Baby had the most efficient planning hull built, but the shock of a seaway was too heavy to continue with this type of flying boat.[38]

When Ens Harold Mott Wilcox arrived in the United Kingdom in February 1918, he was assigned to Killingholme. There he discovered

> *Porte Boats (3 motors, 175 foot wing span)m H-12s, F2as, 7BAs* (sic), *260 and 310 Shorts, Sopwith Schneiders and Pups, Humber* (sic) *Babies, and six Morris Farnums* (sic), *about 300 in all.*[39]

By this time the Baby would have been used only for instruction. Wilcox did not record any flights on the type.

Two Babies were on charge of the RAF on 31 October 1918, while Hallan has recorded the hull of a third was used as the residence for WRNS motor drivers.

> *When the Porte Baby was finally dismantled, her hull was placed in the grounds of a woman's hostel, a door was cut in the side, electric light laid on, and four Wren motor-drivers found sufficient room inside to sling their hammocks, stow clothing, and room even for mirrors and powder puffs.*[40]

Above: A production Baby on the water with the king posts faired in as anti-skid fins and the engines covered. Only Felixstowe and Killingholme could provide hangars large enough to take the Baby.

Below: The circumstances of this photograph are unknown but it is thought to have been taken post-war when the idea of a trans-Atlantic flight was once again on the agenda. Unfortunately the serial is obscured by the crude paintwork.

Felixstowe Baby Specifications

Source	J. Porte's Report 27 July 1916 (1)	Types of British Seaplanes	AIR 1/148/15/76 (2)	J.D. Rennie	Table of Performance of Seaplanes & Flying Boats (3)	JM Bruce Data	Tech Dept Chart NM129 (4)	Tech Dept Chart NM129A (4)
Span, Upper	120 ft	124 ft	125 ft			124 ft	124 ft 0 in	124 ft 0 in
Span, Lower	80 ft							
Length Overall	61 ft	63 ft	60 ft			63 ft	63 ft 0 in	63 ft 0 in
Length Hull						56 ft 10 in		
Height		25 ft	28 ft			25 ft	25 ft 0 in	25 ft 0 in
Chord	12 ft							
Areas, Sq. Ft.								
Wings Total	8,400		7,500			2,364		
Lifting Surface					2,088			
Weights in Lbs.								
Empty					12,892		14,700	14,700
Weight Structure			8,485			14,700 production		
Powerplants			4,140					
Fuel & Oil					3,963			
Fuel & Tanks			4,500 (8 hour)			2,482	2,482	2,482
Load			870					
Military Load					245	298	298	180
Crew					830	1,120	1,120	850
Loaded (Gross)				16,500	17,930	18.600	18,600	16,700
Capacities in Gallons								
Oil			35 (9 hours)					
Fuel			480 (9 hours)					
Engines					3 x 250-hp Rolls Royce Mk.II		3 X RR Eagle VIII	3 x RR Eagle VIII
Speed in Knots					68 (sea level)			
At 2,500 ft							76	73
At 10,000 ft							-	68

Felixstowe Baby Specifications

Source	J. Porte's Report 27 July 1916 (1)	Types of British Seaplanes	AIR 1/148/15/76 (2)	J.D. Rennie	Table of Performance of Seaplanes & Flying Boats (3)	JM Bruce Data	Tech Dept Chart NM129 (4)	Tech Dept Chart NM129A (4)
Climb Time to								
2,500 ft							5.3	4.9
6,500 ft							25.1	20.5
10,000 ft							-	44.0

1. TNA file AIR 1/148/15/76.
2. Notes taken from TNA AIR 1/148/15/76 in J.M. Bruce Collection Box 22, RAF Museum.
3. Air Board Chart. Date of Porte Baby test 19.05.17. Copy in J.M. Bruce Collection Box 26, RAF Museum.
4. *Record of Performances of British Seaplanes 1917–Dec'31 1918*. Copy in TNA AIR1/708/27/11/02. Report NM129 of 12.03.1918 & NM129A of 19.12.1917.

Additional data from the 27 July 1916 Report by J. Porte*:

Crew 4 OR Fuel for 9½ hours & no bombs
Fuel 8 hours
Speed 55 knots Fuel for 4 hours & 2000 lbs of bombs
Bombs 10 x 65-lb.
Climb 3,000' in 1 hour
Armament also given as three Lewis Guns and two 520-lbs bombs or 6-pr Davis Gun for short flights.**

* TNA AIR 1/148/15/76.

** *His Majesty's Naval Aircraft built, building or under repair, March 1918 (Corrected to 27th February)*, Admiralty, Air Department, March 1918.

Chapter 3 Endnotes

1. Hallam, T.D. *The Spider Web*, P.24.
2. Sueter, M.F. *Report on Experiments with Seaplane Hulls, carried out at RNAS Air Station, Felixstowe, during 1915–196 – Presented by Commodore Murray F Sueter, C.B., R.N., Superintendent of Aircraft Construction, Admiralty, to Advisory Committee for Aeronautics.* 01.02.17. P.7. TNA DS.IR23/917.
3. Rennie, J.D. "Some Notes on the Design, Construction and Operation of Flying Boats," The Journal of the Royal Aeronautical Society, Proceedings, Seventh Meeting, 58th Session. 18.01.1923. P.137.
4. *Flight*, 21.09.1916. P.822.
5. Photocopy in Author's collection. No citation but the page is titled "Plate 15. – Types of Curtiss pre-War Flying Boats Having one, Two and Three power plants" and depicts the Model F-Boat, H-12 and Cruiser drawn by AG Hagstrom, NY.
6. In the files relating to the submissions to the Commission on Award to Inventors held after the war, there are many references to the number of Curtiss flying boats built and supplied by Curtiss, but there is no mention of any machine that fits this description.
7. Quoted in letter – J.M. Bruce to Bill Montimer, 07.02.1987. Box 42, JM Bruce Collection, RAF Museum.
8. *Ibid.*
9. This is probably the initial hull length before it was lengthened. The typed text shows 60 ft and this has been altered by Porte to 53 ft. Maj W.F. Vernon, RAF, *Development of Flying Boats*. Report in TNA MUN8/35. Porte corrected and amended this Report of Vernon's.
10. J.M. Bruce thought that it was difficult to imagine that Felixstowe in 1915 had the facilities and staff to build a large boat hull such as that of the Baby. He considered that May, Harding and May may have built the Baby hulls.
11. Maj W.F. Vernon. *Op. Cit.*
12. Porte's notes to Maj W.F. Vernon's, *Development of Flying Boats. Op. Cit.*
13. Cmdr J.C. Porte to CO, Fleixstowe. *Trials of "Porte" Seaplane*, 28.03.16. TNA AIR1/149/15/104.
14. Mines were actually dropped by Short 184 floatplanes in a series of tests. The idea was abandoned but was to be reintroduced in WWII. See TNA AIR1/72/15/9/141.
15. From CO, RNAS Felixstowe. *Duration Test with Porte Flying Boat*, 10.05.16. Signed by JC Porte. TNA Air 1/148/15/76. Copy notes in JM Bruce Collection Box 22, RAF Museum.
16. S.F. Hollister served at Felixstowe and remembered the first Baby but his recollections were that it was powered by Rolls Royce engines. "I remember a "Green' engine being tested in one of the big sheds and casting its propeller which went through the roof." Hollister to J.M. Bruce, 07.02.1956. RAFM JMB Collection Box 42.
17. Porte's Report is dated 22.09.1916, and was included in: *Report on Experiments with Seaplane Hulls, carried out at RNAS Air Station, Felixstowe, during 1915–196 – Presented by Commodore Murray F Sueter, C.B., R.N., Superintendent of Aircraft Construction, Admiralty, to Advisory Committee for Aeronautics*. 01.02.1917. TNA DS.IR23/917.
18. Flt Lt M.J. Day was a pilot of HMS *Vindex*. He was lost off the Belgium coast.
19. CO Felixstowe to DAS. *Experiment of Slipping a Bristol Scout from the top of the "Porte" Flying Boat when in flight*. 08.04.1916. TNA AIR 1/148/15/76, copy notes in JM Bruce Collection Box 22, RAF Museum. See also Jarrett, P. "New Light on an Airlaunch", *Cross & Cockade International*, Vol.19, No.1, P.40.
20. Caption to photograph of the composite Scout and Baby. Probably from *Flight* or *Aeroplane*, article on composite aircraft.
21. Porte's notes to Maj Vernon's, *Development of Flying Boats. Op. Cit.*
22. Copy in TNA AIR 1/148/15/76.
23. Report on Porte Baby 9800. TNA AIR 1/2586.
24. The Report: *His Majesty's Naval Aircraft built, building or under repair, March 1918 (Corrected to 27th February)*. Admiralty, Air Department, March 1918, records the aircraft as the "Porte F.B.2."
25. Sturtivant, R. and Page G., *Royal Navy Aircraft Serials & Units 1911–1919*, P168.
26. *Aircraft & Seaplanes in Use by RNAS* (Undated). TNA AIR1/6A/4/48.
27. Letter of 14.02.1917 marked "Secret" dealing with "policy as regards the RNAS." The letter appears to have been written by the DAS to the CinC. TNA ADM1/8478/10.
28. Ackery, E.M. "Tales of an indifferent naval pilot," *Aeroplane Monthly*, UK, Vol.9 No.12, December 1981, P.682.
29. The other members of the crew were Lt M.W.W. Cross (engineer), Sub Lt H.O. Fry (armament officer), and Air Mechanics H.D.M. Davies and C. Spikings.
30. Hallam, T.D. *Op. Cit.*
31. According to Henshaw (*The Sky Their Battlefield*). P.477, the attackers came from *SFL* 1.
32. Although the boat managed to return to the UK, it was granted as the second victory of Friedrich Christiansen, who flew Hansa-Brandenburg W.12 Marine Number 1183 in the fight. Gutman, J. *Naval Aces of World War 1*, Osprey, 2012. Hobbs was awarded the DSC in June 1917, followed by the DSO the following July, and a bar to his DSC in November. He was the chief instructor on flying boats and trained many British and later USN airmen.
33. "The Hornets of Zeebrugge", *Cross & Cockade* Journal, USA. Vol.11, No.1, 1970.
34. A.B. Massey to Sir A. Robinson, 27.10.1980. Copy in RAFM J.M. Bruce Collection Box 22.
35. The details of the Baby's performance and characteristics recorded here are taken from Maj Vernon's, *Development of Flying Boats. Op. Cit.*
36. Rennie, J.D. *Op. Cit.* P.137.
37. Nicolson, D. "Design and Construction of Flying

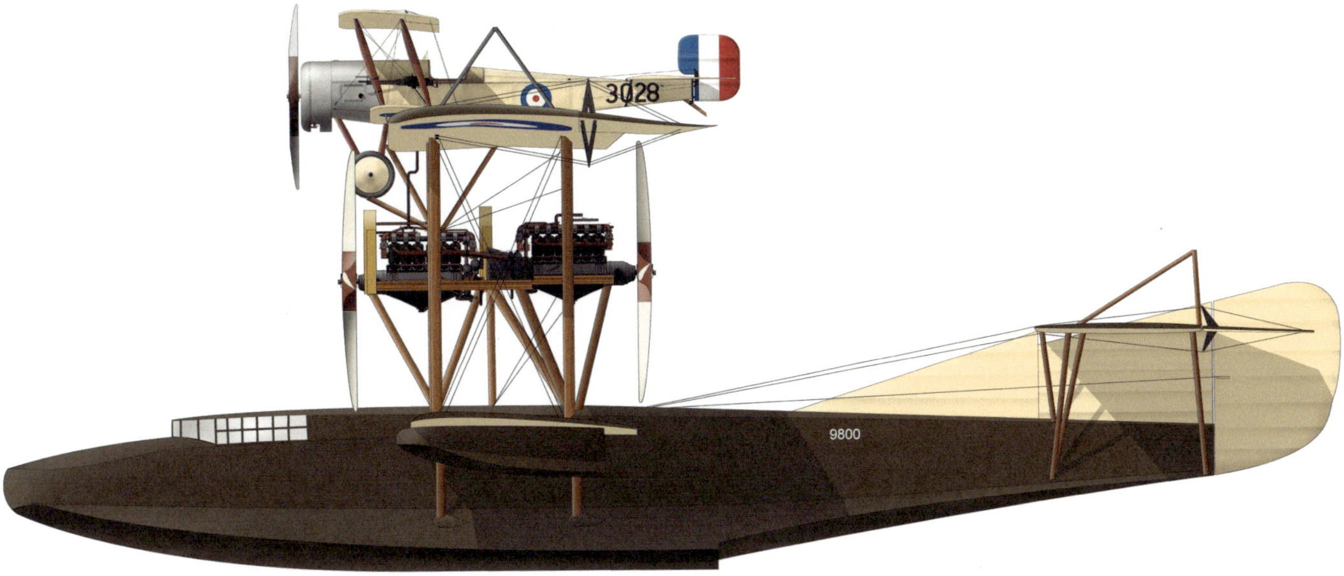

See P.159 for photographs of the Porte Baby and Bristol Scout composite.

Fully armed Baby 9810. This boat was in service with Eagle VIII engines by October 1917 when attacked by three German seaplanes near the North Hinder lightship. She survived and served for a full year before being deleted by the end of 1918.

Boats", paper read before the Institute of Engineers and Shipbuilders in Scotland, reproduced in *Flight* magazine, 03.07.1919. P.874.

38. *Flying Boat Development* – a Report by Maj W.F. Vernon, RAF, reviewed by Col Porte. *Op. Cit.*

39. Knapp, M.W. "Flying Boats over the North Sea," *Over the Front* Journal, Vol.9 No.2, 1994. P.125.

40. Hallam, T.D. *Op. Cit.* P.25.

Experimental Flying Boat Hull 9800

Scale $\frac{3}{16}$" = 1 Foot

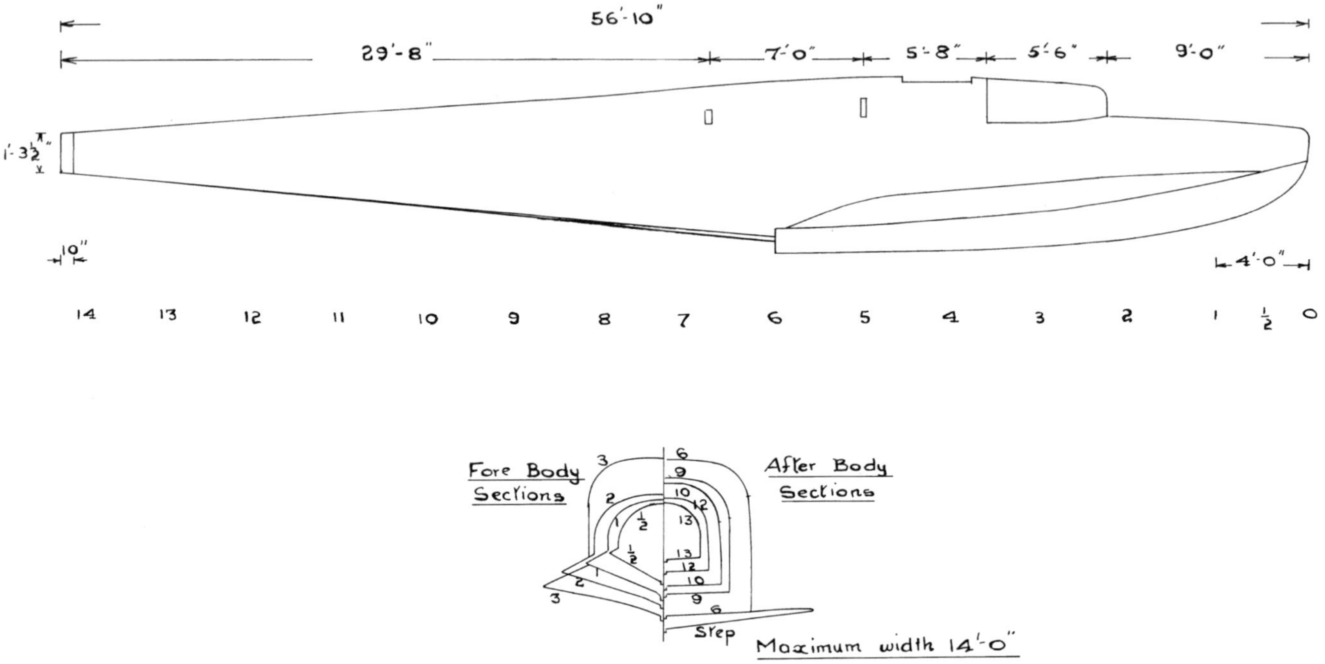

Experimental Flying Boat No 9800

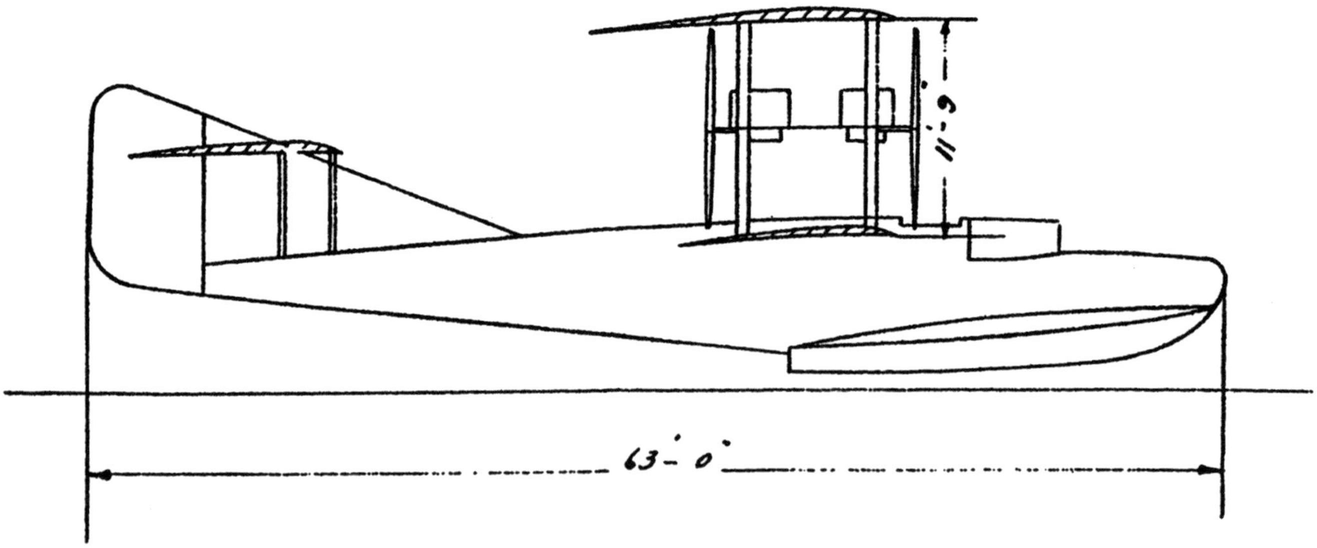

FIG. 19.

Section No 5

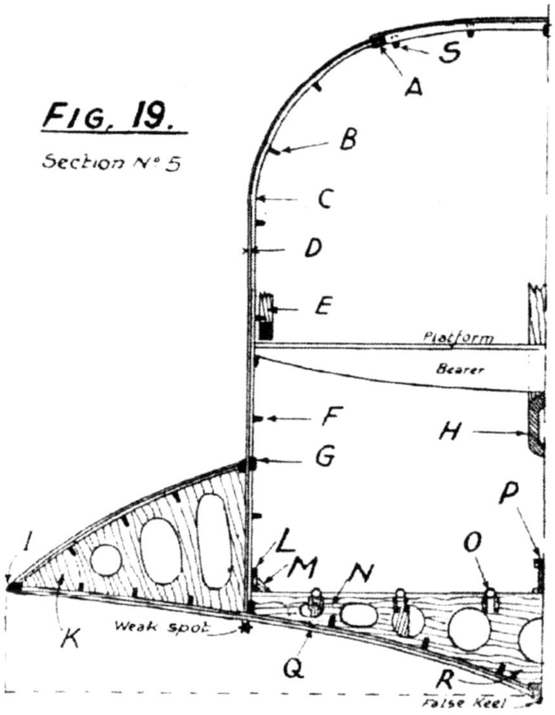

PORTE BABY " AS BUILT 1916 (SECTION No. 5): S, Stringer not continuous with carline of hatchway. A, Carline 1 in. by ¾ in. on flat. B, Six stringers, spruce, 1¼ in. by ¾ in. to ⅝ in. C, Timbers rock elm [illegible] in. by [illegible] in., spaced 2½ ins. D, Planking, inner diagonal [illegible] in. mahogany, outer fore and aft [illegible] in. cedar. F, Diagonal strut, spruce, 1¼ in. by 1¼ in., let into stringers. F, Stringers cut for struts. G, Heavy stringer on flat. H, Diagonal strut. I, Chine, rock elm, 2 ins. by 2 ins. K, [illegible] in. 3-ply. L, Deep spruce sill. M, [illegible] in. 3-ply gusset at alternate floors. N, Floors ⅝ in., spaced 10 ins. O, Three tubes, 1¼ in. 18 g. mild steel; M.R. = 30 cwt. = 6 cwt. on 20 ins.; total weight 96 lbs. P, Girder keelson cut asunder by two struts. Q, Bottom planking two inner diagonal skins [illegible] in. mahogany; one outer skin [illegible] in. mahogany. R, Spruce floors split easily. Weight of boat, 3,000 lbs. Total displacement of hull, 114,200 lbs. Weight of hull per lb. displacement, 0·026 lb. Surplus buoyancy, 444 per cent. Total flying weight, 21,000 lbs. Weight of hull, 14¼ per cent. of total flying weight. Objectionable features in construction: 1. Struts cut away part of stringers. 2. Centre struts cut girder keelson and render it useless in tension. 3. The steel tubes are of little use as keelsons and the fastenings are weak. 4. There is no continuity fore and aft except keel, sides and steel tubes. 5. There is no continuity athwartships to connect fins to hull except skins and timbers. This is a very weak point. 6. The top of fin is in compression, and should therefore be straight, not rounded. 7. There is no proper member to take the thrust of the heels of side struts which are not properly connected to boat sides to distribute the load. 8. The main floors split easily. N.P.L.: The floor broke up when evenly loaded with 796 lbs., supported at ends and centre as in boat.

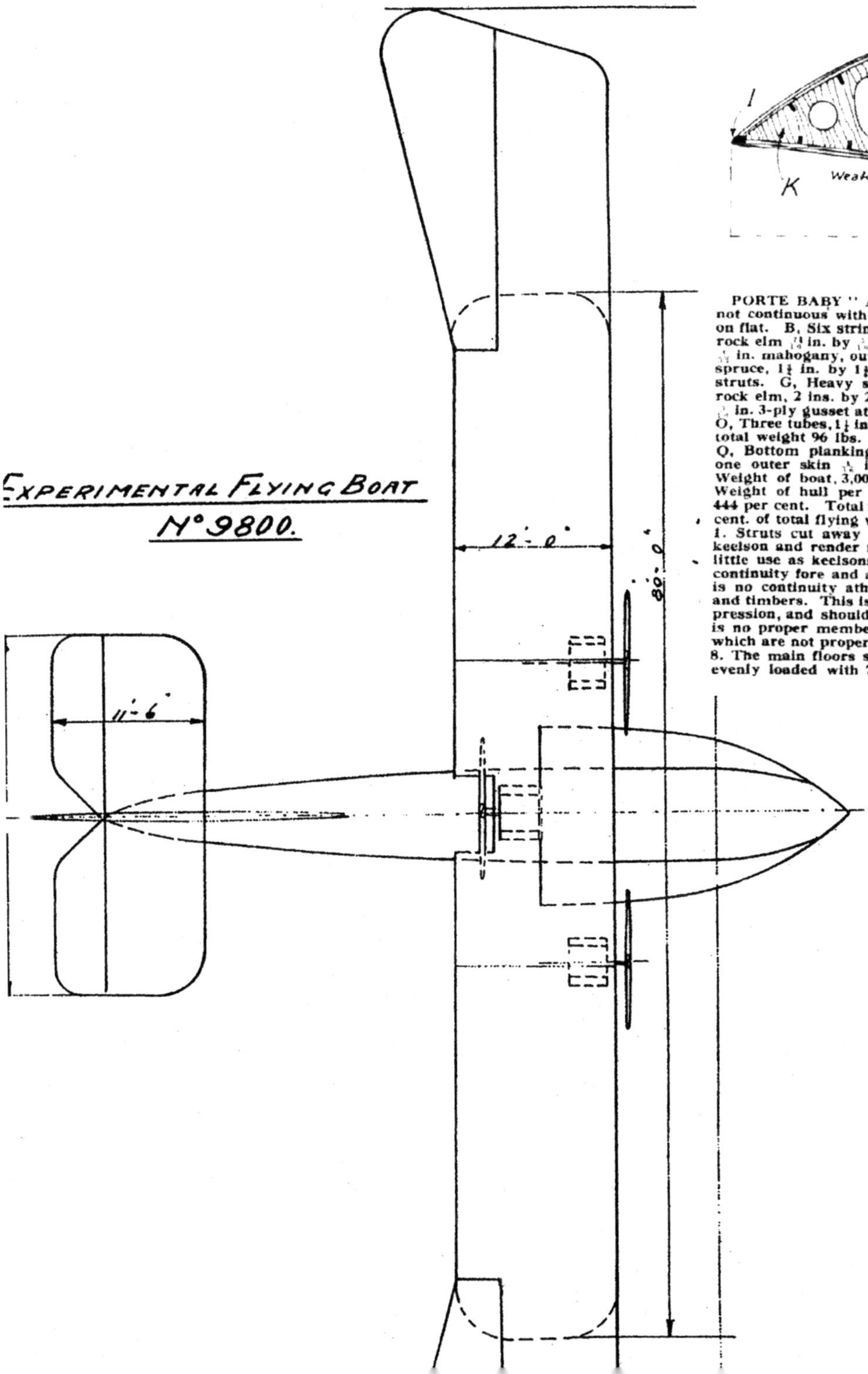

F.2A in dazzle scheme assigned to Great Yarmouth NAS and flown by Flt. Lt. Leckie.

F.2A N4283 in the black and white scheme adopted by G.E. Livock and Flt. Lt. Bob Leckie.

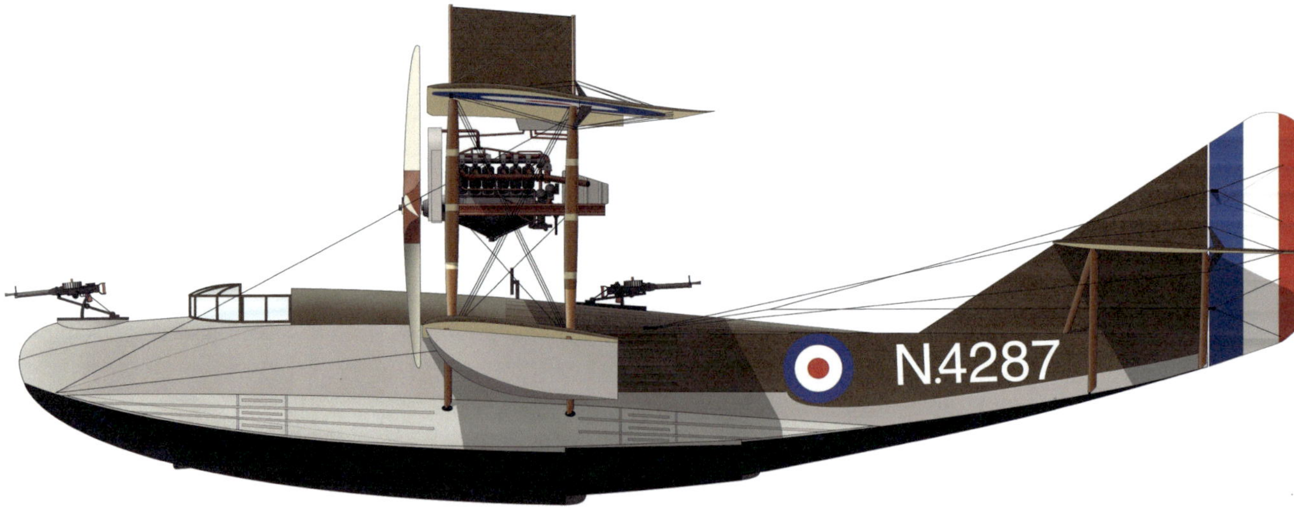

N4287 was a Felixstowe Station boat and was in service form March 1918. It dropped bombs on U-Boats on two occasions and was used by the USN Killingholme by July 1918. Deleted W/E 05 September 1918. Note the unusual large serial application.

4. The America Flying Boats in Combat

Above: The modified H-4 951 with Davis Gun on top wing. In the foreground is standing an officer with a woman and child taking in the work on the slipway, pointing to a casual attitude to the work at Felixstowe at this time.

The Curtiss H-1 and H-4

Little has been written about the activities of the H-4 other than its poor performance. A hint as to what the H-4 were engaged in is given in a secret Memorandum of 7 June 1915. In this Commodore Reginald Y. Tyrwhitt, commander of the Harwich Force, pressed for the use Light Cruisers carrying a large seaplane for attacking Zeppelins.[1] "Experience shows that the present method of attacking Zeppelins with the small Curtiss Seaplane is not a success."[2] Unfortunately the activities of the Curtiss seaplane referred to by Tyrwhitt are unknown. However, during the Zeppelin raid of the night 9/10 August 1915, H-1 950 from Felixstowe took off at 05.30 hours in response to the attack by five Zeppelins. To date no confirmation as to the armament carried on this occasion has been found.

Australia's highest scoring pilot of the war, Robert A. Little, flew the H-4 America boats during the period when he was stationed at Felixstowe. His log book entries give a good idea of the training and use of the Curtiss boats at this period. His first flight was in 3571 on 8 March with McKenzie as pilot. They flew around the harbour and Little recorded in his log book that this was his "First trip in seaplane, not as nice as land machines. Are slow and too big, very slow to answer the controls." With the same boat and pilot he had 10 minutes on the controls at 100 feet on the 16 March. On the 23rd he flew up to 500 feet and was in the air for 60 minutes in 3567. On 1 April he undertook a patrol with Hobbs as his pilot to the Kentish Knoch "Armed with Lewis gun (4 trays) and Revolver (20 rds)." Another patrol to the Wash occurred on the 6th in 3560. The following day he made his first solo in 3571 "Got along alright." He left at 0955 the next day in 3560 with Flt Sub Lt W.H.S. Aplin as passenger, but the patrol did not go so well. The patrol was to the Kentish Knoch with the usual armament referred to above.

> *1035 port engine began to miss. 1040 sighted 4 trawlers so landed beside them. On landing starboard engine moved on its bed, a prop bit side of boat and broke hitting port engine and then going through top plane. Towed to Grain.*

According to the log book, he only made four more flights in the *America* boats conducting "gun practice" in 3560 and 3569, the last such flight occurring on 30th. In between times he flew Short and Wight seaplanes. His last two recorded entries for Felixstowe were on 2 and 3 May when the Wight seaplane "trial did not get off."[3]

At the end of July 1916, five *America* boats were sent out to Malta as German U-boats had passed through the Straits of Gibraltar and were operating in the Mediterranean. Curtiss H-4 3561 was stationed at Malta and is recorded as making a 2 hour 10 minute Mine Patrol on 5 August 1916, and a 1 Hour 55 minute patrol on the next day. Both days there was

Above: The first Curtiss H-12, 8651, on the slipway at Felixstowe. Note the spigot machine gun mount for the front gunner and the cockpit entry opening in the hull behind the early type cockpit canopy. The rudder stripes run the full length of the rudder.

"Nothing to report." Malta also had Short 184 floatplanes and F.B.A flying boats at this time. No submarine intercepts were ordered during the week and only one submarine patrol was carried out. "This has been entirely due to the shortage of machines in commission" and was not a reflection of the capabilities of the H-4.[4] It was proposed to use the surviving *Small Americas* to teach pilots the art of flying the big boats when the *Large Americas* entered service.

The Curtiss H-12

The raids of the German Navy Zeppelins against the UK are well known; however, the airships were also used for mapping

Recognition of Large America & Felixstowe Flying Boats	
H-12 First series	Sharply pointed hull, completely enclosed pilot's cockpit with entry on port side. Oval front gunner's cockpit.
H-12 Improved	Cockpit canopy more square and separated from rear of hull by gap. Scarff ring in front gunner's cockpit.
H-12 Convert	Identical to F.2A. Modifications were incorporated as to what each station had applied to their F.2A boats.
H-12 USN A-794	The first USN H-12 had the early cockpit canopy but no front gunner's position in the bow.
F.2A Early	Porte hull. Cockpit canopy. Note: some converted to late configuration. Small window could be raised to fill gap between cockpit canopy and rear hull.
F.2A Late	No cockpit canopy; balanced ailerons.
H-16 British	Dimensionally same as F.2A, Eagle engines. Gap between cockpit canopy and rear hull.
H-16 USN	Dimensionally same as F.2A, Liberty engines. Gap between cockpit canopy and rear hull.
F.3	Externally similar to F.2A. Cut-away at centre section to allow for rotation of airscrews.
F.5	Externally similar to earlier F boats with open cockpits. Rudder balance in fin.
F-5-L	Externally similar to the F-5 initially. The fin & rudder were progressively modified to a large curved fin with unbalanced rudder.

Above& Below: Curtiss H-12B N4332. Note the different arrangement of the cockpit and bow as compared with 8651. The cockpit canopy now has flat windows and there is a prominent gap between the canopy and the hull to the rear. A circular cockpit with a gun ring has been incorporated for the gun layer. The hull appears to be the same dark colour as the fabric surfaces. This boat was of an order for 24 of the revised design. Arrived at Felixstowe for erection in December 1917. Served at Grain and was patrolling from Great Yarmouth by March 1918. Deleted the following May.

Above: The most famous of the *Large America* flying boats, Curtiss H-12 serial 8666. Points to note are the lack of underwing roundels, the overall light coloured hull, and the rudder stripes not being full length, and style and colour of the serial. The window behind the cockpit was a station modification to let light into the W/T operator's position. The beaching trolley is well marked. Hulls tended to show slight variations and it was usual to keep a trolley separate for an individual aircraft.

mine fields and for long range reconnaissance ahead of the German fleet. As the German High Seas Fleet was smaller than the British Grand Fleet, to have advance warning of the location of the British ships was a decided advantage for the Germans. The year 1916 saw the Battle of Jutland and while the question of whether it was a victory or stalemate will no doubt be fought forever, the influence of air power in the battle can be summed up as being totally ignored by the commanders involved.

The arrival of the Curtiss H-12 came on the scene at the correct moment in the war as the Germans had standardised the production of U-boats and they were being produced in large numbers from their ship building yards in 1917. An unrestricted submarine campaign had been announced by Germany in February with British losses in shipping increasing accordingly. The *Large America* boats gave the British a weapon to use against the U-boats and Zeppelins. Felixstowe developed and operated the Spider Web. This was a patrol zone 60 miles across that allowed for searching of 4,000 square miles of sea and was right across the path that the German U-boats had to take.

The Spider Web Patrol was based on the (Netherlands) *North Hinder light vessel that was used as a central point. It was an octagonal figure with eight radial arms thirty sea-miles in length, and with three sets of circumferential lines joining the arms ten, twenty and thirty miles out from the centre. Eight sectors were thus provided for patrol, and all kinds of combinations could be worked out. As the circumferential lines were ten miles apart, each section of a sector was searched twice on any patrol when there was good visibility.*[5]

The Spider Web covered the North Sea, St George's Channel and the English Channel. The flying boats from Felixstowe would fly out to the Spider Web, fly a radial arm as per their instructions, and return to base.

The other method used was the "Sweep Patrols." In these patrols "machines flew abreast at varying intervals depending on the visibility and the degree of enemy aircraft activity. In the southern part of the North Sea this distance rarely exceeded five miles." The Sweep method covered an area more thoroughly and efficiently when three or more machines were employed, providing that the distance between machines was not too great, enabling them to close on the leader in the event of attacks by enemy machines. "If the sweep could be arranged to embrace land-falls, lighthouses, buoys or any other aids to navigation, its value was greatly increased."

The best height to patrol at was between 800 to 1,000 feet. Any higher and it was not possible to detect a submarine's periscope unless the submarine was moving at high speed, and then only under average sea conditions. At lower altitudes it was not possible to successfully deliver an attack as the submarine could submerge after being spotted, and it was impossible to make use of your height to arrive quickly over the objective. The attack had to be delivered as soon as possible and this "was accomplished by diving full out at the target, and levelling up only in sufficient time to allow the machine to settle on a steady course and set speed." By these tactics the bombs would be dropped at the height compatible with safety and efficiency, never under 300 feet.

Above: 8669 moored out on the water. The opening in the top of the hull under the wing can be seen. It was a difficult position to operate a gun from and the H-12 was virtually defenceless from the rear. Delivered to Felixstowe for erection in February 1917, it was serving with the Felixstowe War Flight by late June. 8669 attacked a submarine on 3 September. Flt Lt D.F. Ellis was injured in an airscrew accident on 9 March 1918. Whether this was a result of the original airscrews supplied by Curtiss is unknown. Moved to training duties it ended up at the Killingholme Seaplane School and was used by the USN. Last recorded in September 1918.

In the North Sea when a submarine totally submerged it was not possible to locate it by the air bubbles from its ballast tanks, therefore the chances of a successful attack more than two minutes after it dived were remote.[6]

The morning of 13 April 1917, saw the inauguration of patrol flights from Felixstowe. H-12 serial 8661, to be affectionately known as *Old '61*, was run down the slipway into the water. The 1st pilot was *Billiken* Hobbs with *Pix* Hallam as 2nd pilot, together with a wireless operator and engineer. They flew the first patrol of the Spider Web. No submarines were spotted but they had proved that the new boats could undertake these long patrols under normal conditions.[7]

Basil Deacon *Billiken* Hobbs from Sault Ste Marie, Ontario, was "short, stocky, and with plenty of energy" and "was one of the best boat pilots in the service." He possessed those qualities that a good boat pilot needed – able to handle his boat under any circumstances, a good navigator, a tireless observer, a man with sea sense and seamanship, good physical stamina or nervous staying power, one who could endure monotony and wait for the correct opportunity and recognise it when it occurred.[8] Many Canadians served in the RNAS and were involved in operations of the *Large America* boats in greater proportion than their numbers would suggest. During the last eight months of 1917, three-quarters of the boat pilots flying from Felixstowe were Canadians. They have left a rich mine of material on the activities of the Naval Air Stations. Hobbs would receive two DSC for attacks on enemy submarines. Post-war German records revealed that there were no U-Boats lost on the days of his attacks. This was quite common for claims made against submarines in the war and reflected the technology of the age. The weapons were not sufficiently powerful to destroy a submarine on their own. Hobbs became Chief Instructor on the *Large America* flying boats for many British, British colonial and US Navy airmen. Hobbs joined the Canadian Air Force in 1920, resigning in 1927 when he established an importing business in Montreal. During his time in the service he made many long-distance survey flights. Recalled to the RCAF in WWII he was commissioned Group Captain and served in Nova Scotia on convoy and anti-submarine patrols. He died in 1963.

Theodore Douglas *Pix* Hallam was also a Canadian, hailing from Toronto. He had learned to fly at the Curtiss Flying School at Hammondsport, NY in early 1914. On the outbreak of war he volunteered his services to the Army and the Navy as an aviator but was turned down as the war would be soon over. In order to get into the war he enlisted in the Canadian Expeditionary Force and was shipped overseas with the first Canadian contingent. He transferred to the RNVR when in the UK. Wounded at Gallipoli he won the DSC. He then went into the RNAS but was not to fly because of his wounds. Using his nickname *Pix* he wrote the classic *The Spider Web* that tells the story of the air war from Felixstowe Naval Air Station. He had been banned from flying because of his wounds but went ahead anyway and flew combat missions. He won the DSC

Above: An early H-12, 8677 at Felixstowe. Note the front gun layer's position with his Lewis Gun on a spigot mount. The officer at the rear stands in the cockpit entry and has another gun mounted at this position. The hull appears to be a light colour above and a dark colour below. This machine was delivered on 15 March 1917.It was credited with shooting down the Zeppelin L43 on 11 June. It is shown here with its original Curtiss hull. Converted with a F.2A hull it served until April 1918 when it was shot down. Note the personnel showing the Lewis gun positions.

twice more on anti-submarine flights. He won a bar to his DSC on 23 April 1917, for an attack on a U-Boat in 8661, but again, the Germans recorded no U-Boats lost on this day. As a Major in the RAF, Hallam flew the last Felixstowe boat, the Fury, with 24 passengers aboard, 5,000 lbs of sand and fuel for seven hours.

On 15 April the War Flight sighted their first submarine, again it was *Old '61* with Hobbs as 1st pilot. Unfortunately the 2nd pilot had not been trained in bomb release and failed to release the bombs. On the second pass he again failed to pull the release levers but pulled the Bowden cables pulling them from their fastenings. As a result the attack was aborted, the U-boat getting away with only a scare. Hallam does not record what the infuriated *Billiken* said to the unfortunate 2nd pilot. Success was achieved on the eighth patrol when Monk Aplin dropped his bombs on the wake of a submerging submarine. Attacks on submarines were to continue. (See Appendix No.4.1).

On 24th April, Calshot reported that Flt Lt C.L. Scott with Flt Sub Lt J. Phillip Paine (observer) and crew attacked an enemy submarine while on patrol from Portland to Calshot while flying *Large America* seaplane 8655. This was one of the H-12 boats of the first batch that had been originally ordered as H-8 boats. Paine reported that their

course was South, submarine heading West and when first sighted was about 1 mile ahead, the visibility being bad. We immediately prepared to attack, and the submarine at once got under way, proceeding West and diving. The pilot of the machine, Flight Lieut. C.L. Scott, carried out my directions for steering immediately over the enemy craft. Two 100 lb. bombs, delay action fuze, were dropped, the first one falling on the enemy's port beam slightly ahead of his conning tower, about 5 feet away from it, and exploded just as the enemy craft became submerged. Our machine was then turned in a very small radius and a second bomb was dropped which exploded about 50 feet West of the first one, in the direction in which the submarine was heading.

The machine remained circling around for about 30 minutes and during the whole time large quantities of oil were coming to the surface. Large bubbles were seen coming up among the oil during the first few minutes, the bubbles subsequently got fewer but were coming up at odd intervals for a long time. The bubbles were at least 6 feet in diameter, being regular upheavals.[9]

The crew thought that they had sunk the enemy, but it resurfaced about an hour later and was attacked with depth charges by a destroyer that then claimed they had sunk the submarine but the U-Boat eventually returned to base.

Above: 8677 inside the Felixstowe hangars showing a bomb loaded under the lower wing.

Patrols were instigated from Fishguard, Land's End, the Scillies and Plymouth. These were supplemented by mid-Channel patrols from Bembridge, Calshot, Portland, Newhaven and Cherbourg.[10]

On the 26th Flt Lt C.L. Scott reported that he left Calshot at 11:10 am (GMT) with Flt Sub Lt F.E. Fraser (observer); AM1 H.G. Renett (engineer) and AM2 C.S. Lacock (W/T operator). At 11.55AM (G.M.T.) a submarine was sighted...on the starboard bow, my course being Easterly. The submarine was awash and stopped, about 2 miles away.

I proceeded to attack, submarine at once diving, and when still about half a mile away, she was practically submerged. One 100lb bomb was dropped over the spot where submarine was judged to be. She had been submerged for about 20 seconds. I then circled round in the vicinity for about 30 minutes, and nothing more was seen of the submarine; I then proceeded on patrol and remained away from the position for about 2 hours with the object of giving the submarine time to re-surface.

At 3.47PM I had returned to the vicinity and again sighted the submarine...about 2 miles on my port bow. My course being Westerly, I proceeded to attack and the submarine again submerged. Two 100lb bombs were dropped, the first of which was seen to explode directly over the submarine, as indicated by the wake which was still visible.

I am of the opinion that as a delay fuze was fitted to the bombs it is most probable that the submarine or its periscope were damaged.

I was then forced to return to Calshot owing to lack of petrol.

The observer and remainder of crew carried out their duties with great coolness.[11]

The CO of Calshot wrote that he considered that *this officer and the other members of the crew carried out this attack in a very creditable manner and I think great praise is due to Flight Lieut. Scott as this is the second occasion within four days that he has succeeded in attacking a submarine with this machine.*[12]

The first day of May saw *Perham* and *Tiny*[13] from Felixstowe attacked by a twin-engine German floatplane. Encounters between the boats and the floatplanes were to continue eventually seeing large numbers of both engaged in aerial combat as the fight for aerial supremacy in the North Sea grew in intensity in 1918. The *Tiny* referred to by Hallam was Canadian John Osborne Galpin, so named because of his generous proportions. Not to be confused with the British Capt C.J. Galpin who was also a big boat pilot, *Tiny* Galpin suffered for a string of bad luck being denied successful attacks on U-Boats by mechanical failures such as that on 22 December 1917, when the bombs failed to explode. He became comfortable with being let down by his engines and spending a night at sea. On one occasion the destroyer that was to tow him home rammed the flying boat, sinking it instead. He was so depressed by his bad luck that his squadron mates gave him a pebble in a small silk bag with the tale that it was an Egyptian talisman, and thereafter his luck changed.

The 13th of April was also an important day for the Great Yarmouth NAS as they received their first H-12 serial 8660 flying boat on that day. It was to be 1 May before the first patrol was flown. On 5 May 8666 arrived at the station. This boat was to go on to become the most famous flying boat of the RNAS during the war. Her illustrious career commenced on the 14th, only nine days after her arrival at Yarmouth.

With a crew of Flt Lt Christopher J. Galpin (1st pilot in command), Flt Sub Lt Robert *Bob* Leckie[14] (2nd pilot), CPO V.F. Whatling (W/T operator), and AM J.R. Laycock (engineer), 8666 left at 3.30AM armed with incendiary ammunition on an anti-Zeppelin patrol. A Zeppelin was spotted at 3,000 feet below the boat that was at an altitude of 6,000 feet. They dropped their bombs to lighten the boat and Leckie took over the controls. Galpin moved to the forward cockpit to the twin Lewis guns mounted parallel in the bow. Diving at the airship Galpin opened fire and within a short time the whole airship was a glowing mass that fell into the sea. The airship was *L.22* and she had been looking for mines off Terschelling.

After returning to base and reporting their success, their Squadron Commander wrote that he considered the conduct

Above: 8677 on the Felixstowe slipway. Note the camouflage of the hangars in the background.

of the officers and crew were "deserving of the highest praise considering the long flight there and back, their successful navigation, the unfavourable weather conditions, the weather being thick and hazy with frequent rain, and their determined and successful attack."[15] Galpin was awarded the DSO, Leckie, the DSC, Whatling and Laycock the DSM. The letter to their Squadron Commander noted that these "honours will be gazetted, but particulars of the service will not be mentioned."[16] Importantly, the Germans did not realise that the British had a new potent weapon and attributed the loss of *L.22* to thunderstorms that prevented the airship from ascending to a safe altitude and she was brought down by fire from British warships. 8666 was to have six subsequent encounters with Zeppelins.

In June the *Large Americas* were "proving very successful for patrol work and the only difficulty that was being experienced was in teaching new pilots to fly these machines owing to the fact that no dual control was fitted. It was hoped that dual control would be fitted in all the newer machines now on order." Longmore also reported that one machine was being tested with the engines turning right hand rather than having opposite turning engines "as has always been the case in the past." With a larger tailplane and slightly more incidence on one wing the boat was quite comfortable in the air. If successful this would "greatly relieve the position as regards the production of these engines."[17]

The Germans were still unaware that a new weapon had entered the war when, on 14 June, Felixstowe also had success. Five Zeppelins took off to operate in the U-boat blockade area to cover minesweepers dealing with a British minefield 40 miles north of Terschelling. British radio intercepts saw *Large America* 8677, with crew of Flt Sub Lt B. *Billiken* Hobbs, with another Canadian, Flt Sub Lt R.F.L. Dickey[18], as pilots, AM2 H.M. Davis (W/T operator), and AM (E) A.W. Goody (engineer), leave at 5.15AM and fly to near the Dutch coast at Vlieland. At 8.40 they were again off Vlieland and sighted a Zeppelin about five miles off their starboard bow at about 1,500 feet. 8677 was at 500 feet and immediately climbed to 2,000 feet to attack. Hobbs was piloting, Dickey manned the bow gun, Davis manned the midship gun and Goody the stern gun.[19]

As we approached the Zeppelin we dived for her tail at about 100 knots. Her number L.43 was observed on the tail and bow, also Maltese Cross in black circle. Midship gun opened fire with tracer ammunition and when about 100 feet above Sub-Lieut Dickie (sic) *opened fire with Brock and Pomeroy ammunition, as the machine passed diagonally over the tail from starboard to port. After two bursts the Zeppelin burst into flames. Cutting off engines we turned sharply to starboard and passed over her again; she was by this time completely enveloped in flames and falling fast. Three men were observed to fall out of her on her way down. Flames and black smoke were observed for some time after wreckage reached the water. We set course for Felixstowe arriving at 11.15AM.*[20]

So ended the three month service life of the *L.43* with the loss of all her crew.

As a result of this action the Germans ordered a minimum altitude of 13,000 feet for its Zeppelins, thus reducing their effectiveness. The British fleet was harboured at Scapa Flow and if the German heavy ships sortied for a raid on the British coast then it took some time for the fleet to come down from Scapa Flow. British submarines were positioned on the exits from the mine fields protecting the Bay of Filygoland. They were to give warning if the Germans sortied; however, with the Zeppelins able to cruise at around 1,000 feet, the submarines were in a difficult position as they could not come to the surface to carry out their duty. Now that the minimum height of their patrols

Above: The serial on 8681 is applied to a dark rectangle on the starboard side of the hull, and the bottom colour is very dark and fresh. 8681 undertook experiments with towed lighters from Calshot in July 1917. This machine lasted until January 1918 being deleted after suffering a forced alighting on the 14th of that month.

had been raised the Zeppelins could no longer harry the British submarines. Also the airship crews suffered as they described these patrol flights as strenuous as a flight against England.

Not all operations against the Zeppelins were successful. 8666 was on a reconnaissance patrol on the morning of 24 May 1917, when, after about two and a quarter hours, she was at 1,200 feet owing to the low visibility and

a super-Zeppelin with three cars and five propellers suddenly appeared out of the cloud at 2000 feet coming towards us about one mile away; on seeing us he dropped two white flares, we did not answer this signal but put on full speed, dropped our bombs and climbed up at him. He then turned quickly through 16 points and started to climb hard. When he reached 3000 feet we had gained on him and were actually 300 yards astern. He threw out a smoke screen (probably a discharge of water ballast. Author's note.) *under cover of which he gained the main bank of clouds; it was not feasible for us to follow him there. As he disappeared I fired half a tray of Brock, Pomeroy and Tracer into him, but was unable to observe the effect.*

I would point out that we were easily overhauling him on climb, and had the clouds been a thousand feet higher would undoubtedly have made certain of him.

We then climbed through open spaces until we reached the top of the cloud at 10,000 feet, the boat being under light load, and continued W.S.W. until 8.40AM. Without sighting any more Zeppelins, when no land being yet in sight and petrol being very low we descended near some trawlers to enquire our position. We found this to be Cromer Knoll, and we were taken in tow by H.M. trawler Curvia *who transferred us to H.M. Trawler* Rialto *which towed us towards Yarmouth. We were later transferred to* P.25 *who brought us into Yarmouth.*[21]

The *P.25* was a P-Class patrol boat made to represent a German submarine travelling on the surface. The hope was to lure a German submarine in close such that the patrol boat could attack.

8666 continued being active on anti-submarine and anti-Zeppelin patrols. In August 1917 two De Havilland D.H.4 land machines arrived at Bacton, the landplane satellite for Great Yarmouth. These machines with extra large fuel tanks for a 14 hour flight as they were originally planned to be used for a reconnaissance of the Kiel Canal but this never took place. The D.H.4 was a single engined, two-seat bomber that was a fast and high climbing biplane. With the bomb gear removed and flotation bags fitted and extra tankage the D.H.4 had an endurance of six to seven hours. The Zeppelins now operated at an altitude that the *Large America* boats could not reach, and so the D.H.4 would go on an anti-airship patrol with an accompanying flying boat to undertake the task of accurate navigation and to be available to effect a rescue should the landplane have to come down in the sea. On 4 September 1917, 8666 undertook such a task accompanying a D.H.4 in a hunt for Zeppelins. The patrol was aborted due to fog. The next day special D.H.4 serial A7459[22] flown by Flt Lt A.H.H. Gilligan, with Flt Lt G.S. Trewin as his gunner in the rear cockpit, was accompanied by 8666 with Flt Lt Robert Leckie (pilot), AM Thompson (W/T operator) and AM Walker (engineer), and with Sqn Commander V. Nicholl in overall command.[23]

Nicholl reported that around noon when the pair was about 30 miles from Terschelling

A Zeppelin was sighted 25 miles North West Terschelling Island, steering S.E. I altered course to intercept the Zeppelin, H.12 9,000 feet, D.H.4, 10,000 feet, and Zeppelin 10,000 feet. I signalled the D.H.4 to climb as high as possible to attack the

Above: This De Havilland D.H.4, A7830, is readily recognised by its striking black and white colour scheme. Allocated to Great Yarmouth for special service in December 1917, it attacked a U-Boat on 21 March 1918, dropping two 65-lb bombs. The pilot on this occasion was the redoubtable Wing Cmdr Charles Rumney Samson with AM Radcliffe in the rear cockpit. This biplane survived into the 1920s. There is a Fairey Hamble Baby and what appears to be a standard Sopwith Baby sharing the hard stand with A7830. A twin-engine flying boat is in the far background. C.F. Snowden Gamble considered the D.H.4 one of the most successful two-seaters issued to Great Yarmouth in 1917.

Zeppelin.

At 12.30PM, I opened fire on Zeppelin, our altitude was 12,000 feet and Zeppelin 14,000 feet. She dropped water ballast, and climbed still higher. The Zeppelin number was L 44. *I continued attacking unsuccessfully for one hour, firing 400 rounds of Anti-Zeppelin ammunition. The tracers were seen hitting the Zeppelin. During most of this time we were subject to a heavy machine gun fire from the Zeppelin. The Zeppelin, in the meantime, led me over a squadron of two light cruisers and four destroyers, who did not open fire, presumably on account of the proximity of the Zeppelin. The D.H.4 was some distance away, and endeavouring to climb higher and at 1.30PM signalled me that his engine was not pulling well, and he could not climb any higher than 14,000 feet. I then signalled him to close and attack the Zeppelin, which he did without result. At 2.0PM the D.H.4 signalled me that he had serious engine trouble and I signalled him to follow me and we would attempt to make Yarmouth. On passing Vlieland on a course 255° another squadron of two light cruisers and four destroyers were sighted, who heavily shelled us, making very good shooting, fragments of shrapnel damaging the starboard wing. A large group of mine sweepers was sighted at Texel Island.*

While attacking the L 44, *a second Zeppelin was sighted about 10 miles to the Northward, steering a S.Easterly course at an altitude of approximately 10,000 feet. She made off for Borkim* (sic) *Island at high speed, and did not attempt to assist the* L 44.

After the war it was learned that the other airship was the *L 46* and she had seen the attacking aircraft. "The attack was well planned, in so far that the machine had the sun behind it, and therefore could only be seen with difficulty by those aboard *L.44*."[24] *L 46* immediately warned *L 44* by wireless of their danger and that airship dumped ballast to ascend to a safe height.

At 3.30PM when about 50 miles E by N of Yarmouth, the D.H.4 engine failed completely, and the machine crashed into the sea. I immediately landed in the H 12 alongside the D.H.4, and with difficulty, on account of the sea, picked up the crew.

Above & Below: De Havilland D.H.4 A7457 and A7459 were modified for a long range reconnaissance of the Kiel Canal. When this did not eventuate A7459 was chosen as an anti-Zeppelin fighter. Still bearing its fawn and blue camouflage A7459 is illustrated here in the configuration it is thought to have been in when it attacked Zeppelin L.44. Twin Lewis guns are mounted above the centre section in addition to the pilot's synchronised Vickers gun, and the gunner also had twin Lewis guns on a ring mounting. Flt Lt A.H.H. Gilligan (pilot) with Flt Lt G.S. Trewin (gunner) are occupying the cockpits. The light coloured rectangle under the cockpit is believed to bear the aircraft's nickname "ALLO LÉDÉ BIRD." The unusual cockades adopted for this colour scheme are noteworthy as is the application of the camouflage to the lower surfaces.

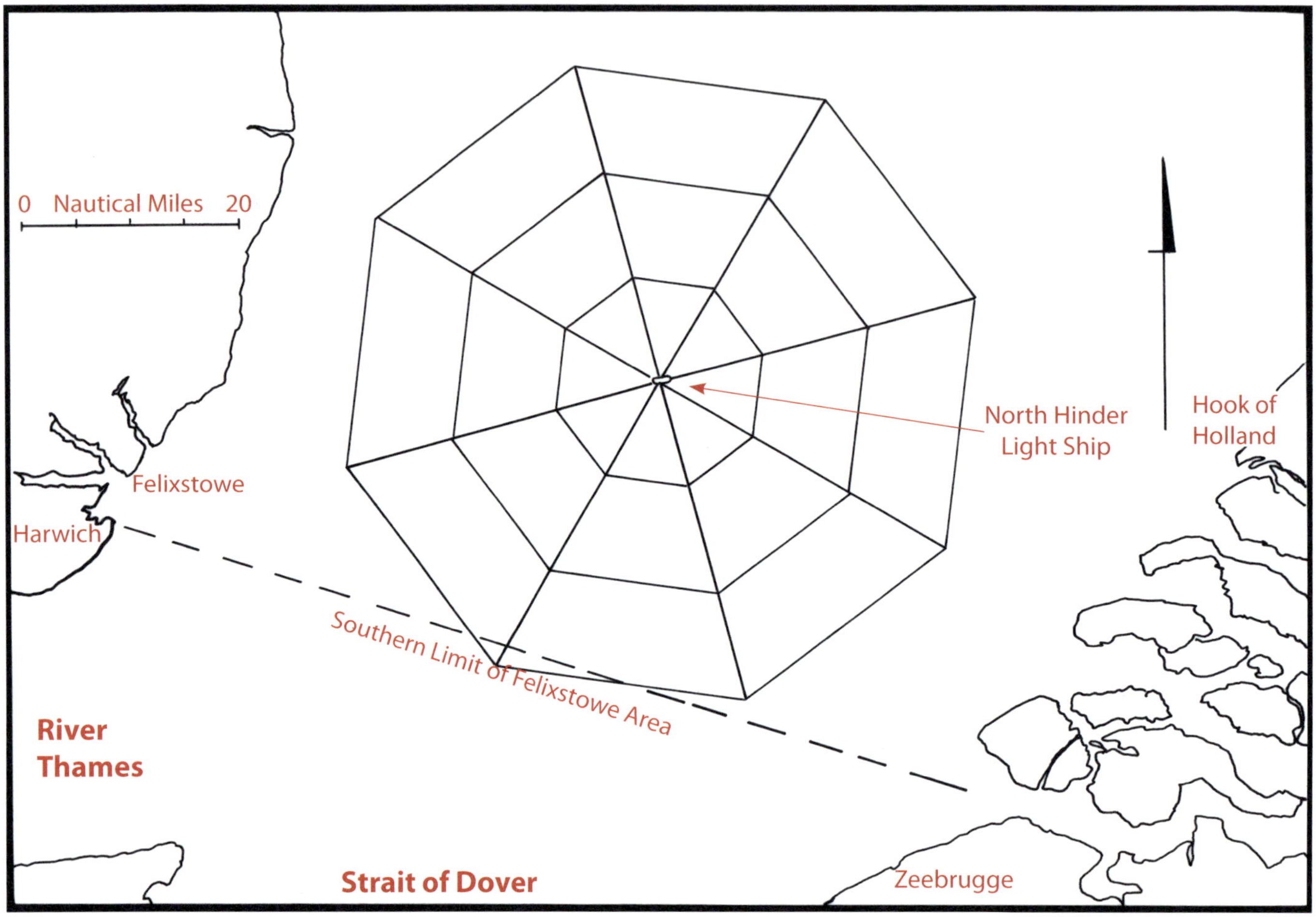

Above: The Spider Web. The anti-submarine patrols were centred on the North Hinder Light Vessel.

Above: De Havilland D.H.4 A7830 shows its unusual camouflage scheme including camouflage on the lower surfaces.

DH.4 A7459 was one of two modified for a reconnaissance of the Kiel Canal. This mission was never flown and A7459 attacked Zeppelin L.44 on 4 Sept. 1917 in company with H-12 8666. A7459 was damaged and forced to ditch, the crew of F/Lt. A.H.H. Gilligan and L/Lt. G. Trewin being rescued by H-12 8666.

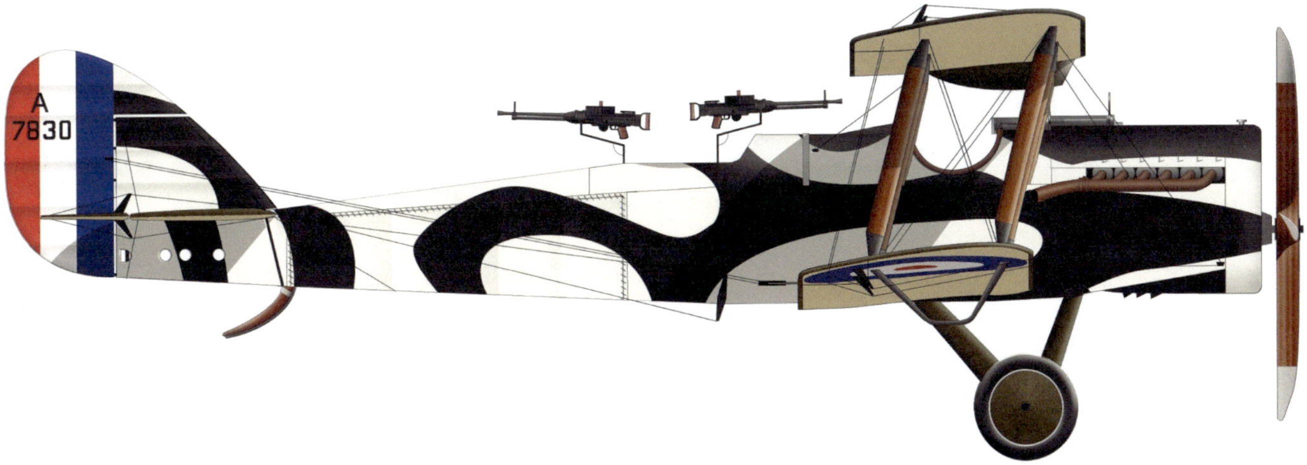

Flown by W/Cdr. Charles Rumney Samson and AM Radcliffe, DH.4 A7830 attacked a U-Boat with two 65 pound bombs on 21 March 1918. A7830 survived into the 1920s.

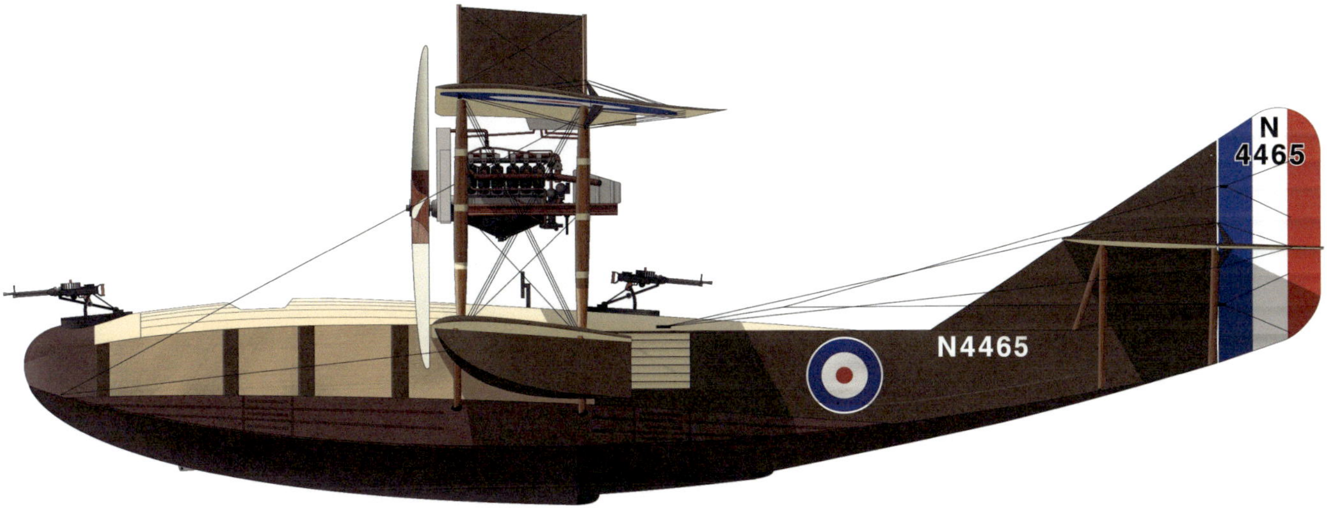

Late F.2A N4465 in one of the "standard schemes" used by Saunders. This boat had balanced ailerons.

Above: 8683 at Felixstowe with the original Curtiss hull. The cockpit entry opening and Lewis gun mount is well shown. In this photograph the hull appears to be the one colour overall with the serial marked on a dark (black) rectangle. 8683 was converted by April 1918. It force landed on land on 30 June 1918, the crew being slightly injured. It was deleted in W/E 18 July 1918. Note the covered Lewis guns.

The D.H.4 sunk almost at once. As the state of the sea made it impossible to get off again and the port engine was not giving its revolutions, I taxied towards Yarmouth till 7.0PM, when I ran out of petrol.

8666 began to fill with water as she had a hole near the step caused by anti-aircraft fire. Bailing was started with empty petrol tins hastily converted to bailing buckets. The short, steep sea lifted the tail up, pushing the bow down allowing water to pour into the cockpit. Leckie had to change course and hoped to meet the War Channel near Cromer but the fuel gave out and the boat was adrift. During the night the starboard wing tip float carried away and the men took turns from bailing by "resting" for two hours on the port wing to keep the starboard wing out of the water.

Two of their four pigeons were released, the message from one reaching Yarmouth on Saturday 8th. On Wednesday evening fuel ran out and the boat now drifted at the mercy of the sea. The boat continued to drift. Another pigeon was released on the 6th but was never seen again. On the afternoon of the same day, the last bird was released. This pigeon arrived at the air station at 10:45 am the next day. This proved that the crews were still alive and the rescue attempts were continued, but in the wrong locations. It was discovery of the message of the 5th that was to lead to the rescue of the missing aviators. The pigeon that had been released on the 5th had fought its way to shore but had died. Its body was found by chance by an office of the 4th Battalion Monmouthshire Regiment on the beach at Walcot and the message passed on. This message led to the search moving to the north. It read:

GOVERNMENT PIGEON SERVICE
H-12 No. 8666, Sept 5th, 4.00PM.
We have landed to pick up DH4 crew, about 50E by N of Yarmouth. Sea too rough to get off. Will you please send for us, as soon as possible as boat is leaking. We are taxiing W by SS.
V Nicholl

At 12:42PM on Saturday 8th, the old gunboat, HMS *Halcyon*, whose task was patrolling the Norfolk coast, found the missing' boat with the two crews still alive. In the words of one of the D.H.4 crew:

With H.12 proceeded to Terschelling: in action with Zepp L.44 *and four squadrons of light cruisers, destroyers and mine-sweepers; fought Zepp for 1½ hours and obtained photos and naval reconnaissance. On return journey engine seized up and we finally crashed into sea at 3PM. H.12 picked us up and taxied until petrol ran out at 9PM. We then drifted about the ocean until 2PM on the 8th when* Haleyon (sic) *picked us up. H.12 leaked and continual bailing was necessary; floats got water logged and we had to take turns in laying down out on*

Above: The location where this photograph of 8683 was taken is not known but has been suggested as in France. It has the original Curtiss hull and shows how the after-body was submerged with the rudder almost touching the water.

wing tips. We had no food the whole time and our water ration was a wineglass full each diem.[25]

The men were on their last resources when rescued. Having no food and only two gallons of fresh water, they had to drain rusty water from the radiators to slake their thirst. Despite the pounding she had received 8666 was salvaged, made serviceable again, and returned to Zeppelin hunting. The pigeon was stuffed and displayed in the Yarmouth Mess above the inscription: "A very gallant gentleman." He now resides in the RAF Museum.

After the men were rescued an analysis was made of the patrol. It was considered that "further tests should be made with the engine of the other D.H.4, and see is these are satisfactory, that the experiment should be made again."[26] Attached to this memorandum was the following:

It will be seen that complete success was only frustrated by a very small margin. Had the engine of the D.H.4 lasted another hour, which it should certainly have been expected to do, I think destruction of one, or both Zeppelins, would have been accomplished.

The ammunition again seems to have been unsuitable for the purpose for which it is intended.[27]

There were discussions on the suitability of the ammunition used following on from these reports. The ammunition used was a mixture of Brock, Pomeroy and Buckingham, "and it appears that the range was too long for these bullets to function." Every effort was being made to increase the range of the ammunition. It was also proposed that "a .45 double barrelled express rifle should be carried in each large America and trials of some Brock ammunition should give a greatly increased range are nearly completed."[28] No documentation of such a rifle being carried by any *Large America* has been found.

Four Rolls Royce Eagle VIII engines were sent to Yarmouth to be fitted into two Large America boats

and it is hoped that these machines will then be able to attain the altitude of the Zeppelins.

In the meantime, further tests should be carried out with the engine of De H.4 machine No.A.7457 and, when ready, this machine together with the two large America flying boats should be sent out on anti-Zeppelin patrols as suitable opportunities present themselves.

Two additional De H.4 machines, Nos. N.6395 and N.6396, have been allocated to Yarmouth for these patrols.[29]

Another incident involved Curtiss H-12B N4340 that was forced to alight in the Channel while on patrol on 14 May 1918. The following day, while on "Z" patrol in Short 184 N1795, Lt Hope, with Wireless Operator and Observer A.E. Ingledrew, saw N4340 on the water. Hope landed and

attempted to convey my emergency rations to her, as she was in need of food. I attached these to a life-belt, and dropped them as near her as I could. Lieut. Oldridge swam out and reached them but was seized with cramp; I picked him up with my wing tip, and having let go my bombs proceeded to Cherbourg....landing at 1850.

Seaplane 4340 had caught fire in the Port engine, and the flames had spread to the whole of the lower plane, which was badly burnt.[30]

This Page & Facing Page: 8666 when found by HMS *Halycon*. The crew took turns on the port wing to keep the broken starboard wing out of the water.

Hope had picked up one of the crew and flew him to Cherbourg. Immediately on arrival of N1795 with the location of the H-12B N4340, the Senior Patrol Officer, RAF Cherbourg proceeded at 1910 in Wight Seaplane 9854, without bombs or passenger, towards the position that Hope had reported finding the *Large America* boat. At 1950 he sighted the H-12B on the water with seaplane F.2A N4543 circling over it and alighted and taxied close to the damaged machine. The sea was choppy and he had no way of re-starting the engine single handed as the starting magneto was in the pilot's seat, away from the starting handle. By flying alone he had hoped to save the crew if they should be compelled to take to the water. Since the situation did not warrant such action he did not attempt to go alongside N4340 but took off and returned to Cherbourg, alighting at 2140 by the light of a petrol flare on a bouy. As he left he saw N4535 alight and take off the remaining crew. N4535 landed outside the harbour at 2200 by the light of French searchlights. The flying boat was then picked up by the Station Motor Boat from outside the harbour in the dark and towed to safety through the boom under very difficult circumstances.[31]

It appears that N4340 may have been salved as it is recorded at Calshot on 25 May and not deleted until the week ending 29 August 1918.

The H-12 flying boats of Felixstowe and Yarmouth did not have all the combats with Zeppelin to themselves, for on 10 May 1918, radio intercepts had given a fair idea of the course of a Zeppelin and F.2A serial N4291, *Old Blackeye*,[32] left Killingholme at 13.20 in "search of hostile airships." The

Above: This H-12 from Calshot identifies its Station by the large white "C" on wings and fin.

crew comprised Capt T.C. Pat Pattinson (1st pilot), Capt A.H. Munday (2nd pilot)[33], AAM Johnson (W/T operator); Sgt H.R. Stubbington (engineer). At 16.30 they sighted an airship on the port beam at an altitude of 8,000 feet. The Zeppelin

was about 1,500 feet above our machine and proceeding due east in the direction of Heligoland. I pointed it out to Captain Pattinson and he at once put the machine at its best climbing angle. I went to the forward cockpit and tested the gun and tried the mounting, and found everything worked satisfactorily. The engineer rating immediately proceeded to the rear gun cockpit and tested both port and starboard guns. At this time our machine was at a height of 6,000 feet. The hostile craft had evidently seen us first, and was endeavouring to get directly over us in order to attack us with bombs. When at a height of 8,000 feet and the Zeppelin had climbed to a height of 9,000 feet I opened the attack and fired 125 rounds of explosive ammunition. The engineer rating also opened fire and fired about the same number of rounds. The Zeppelin was about 500 yards distant. All our fire appeared to hit the craft and little spurts of flam, our explosive bullets, appeared all over the envelope. I noticed much, what appeared to be water ballast, and many objects thrown overboard and then the nose of the hostile craft went up a few degrees from the vertical, this was, apparently checked by the occupants as the craft righted and commenced to climb as much as possible. I had a gun stoppage and spent many minutes clearing it, being obliged to take the breech of the gun to pieces.

The Zeppelin continued to climb and gain on us slightly and again endeavoured to get directly over our machine. The enemy succeeded and dropped five or six bombs. The engineer rating reported to the first pilot (Captain Pattinson) that seven hostile destroyers were circling beneath us. We were now at a height of 11,000 feet and the Zeppelin's height was approximately 12,500 feet. I opened fire again and fired another 130 rounds of explosive and tracer bullets. I noticed the propeller of Zeppelin's port engine almost stop and the craft suddenly steered hard to port. I concluded that the port engine had been hit by our gunfire as well as other parts of the craft, as the envelope and gondolas seemed a background for all the flashes of the explosive and tracer bullets. There was much outpouring of ballast and articles and considerable smoke. I concluded that

Above: The circumstances that led to H-12 Convert 8683 attempting to alight on land on 30 June 1918, are unknown. The crew were all slightly hurt. At Dunkirk by 11 July 1917, this boat apparently went back and forth between Dunkirk and Felixstowe. It was converted with an F.2A hull at Felixstowe by 6 April 1918. Sir Austin Robinson stated that most conversions, if not all, were done at the station where the boat was based. It was written off after the incident shown here. The purpose of the lettering on the white panel under the lower wing is unknown as it does not appear to relate to the serial number.

Above: Curtiss H-12B N4350 was one of the British H-12 boats assigned to Killingholme where it was used by the USN where they flew joint missions with the RNAS, or rather the RAF, as it became after April 1918. N4530 bombed a U-Boat on 8 June with Capt A.H. Munday, RAF, and Ens R.U. Mill, USN, as pilots in command. Two days later it bounced on landing and broke its back. It sank under tow and the result is shown here. The crew at the time were Lt H.A. Madge, RAF, Ens W. Jackson and 2AM P.N. Payne, all were slightly injured. The style of the serial is most interesting and the boat appears to have been a uniform colour to the hull. The men with the wreck are in USN uniforms. (Ens Raymond L Atwood collection/ Emil Buehler Library)

Above: First of several views of Felixstowe Naval Air station showing the large hangars, wooden slipways, flying boats on the hard-standing and overall size of the complex that was overseen by John Porte.

we had finished the Zeppelin and informed Captain Pattison that we had bagged it. But the craft again headed for Holstein in a crabwise fashion emitting much smoke. I had other gun stoppages and one bad jamb. One of the explosive bullets exploded in the gun and flashed into my face and on to my hand but outside of a few scratches I received no injury. Our port engine commenced to give slight trouble and the engineer investigated. He reported that an oil feed pipe had broken and it would not be long before it would break in two. As soon as I had my gun cleared of the trouble caused by the exploding bullet I signalled to Captain Pattinson that I was ready to re-attack and looked for the enemy craft. I saw it still proceeding due east in a crabwise fashion. It was losing height and emitting smoke which was of a black variety though parts of it was pure white. The hostile destroyers underneath opened fire on us and owing to engine trouble and lack of petrol for our return journey, and being only about sixty miles off Heligoland we gave up the attack at 5.35, having been attacking for one hour and five minutes.

The oil pipe broke in tow and we were forced to glide and land in a rough sea. The engineer quickly climbed out on to the top of the engine and repaired the break with tape. Fifteen minutes later we took off. As the machine was difficult to control Captain Pattison asked me to get out and endeavour to ascertain whether we had damaged the rudder or elevator in taking off and in looking back I observed the German destroyers steaming with all possible speed in our direction.

As regards gun jambs:
About 500 rounds were fired. Two and half double pans of explosive ammunition were fired from two guns in the after cabin. No jambs occurred, but the gun stopped twice to misfires.

Two double pans of explosive and half a pan of Mark VII and tracer were fired from the forward guns in long bursts. One serious jamb occurred owing to the premature explosion of a round of Brock ammunition. This premature explosion may have been partly caused by overheating of the gun or it may have been due to a faulty round. The blunt nose of the Brock bullet is liable to cause faults in the feed, and if the sensitive nose of the bullet takes up against the forward end of the feed arm slot may easily premature explosion.[34]

The F.2A often had trouble with its fuel system due to the great length of piping involved. The main tanks were in the hull and fuel was forced up to the gravity tanks in the upper wing centre-section by wind-driven pumps, the engine carburettors being fed from the gravity tanks.

The crew were given credit for the destruction of *L62* which fell that day. Actually they had been attacking *L56* and had never placed the airship in jeopardy. *L62* was seen to enter clouds by surface craft and then falling in halves, one nearly landing on the trawler *Bergedorf*. Lightning was thought to have been the cause of the disaster. The *L62* was the third of the V-class Zeppelins. It had first flown on 19 January 1918.

According to T.C. Pattinson "Poor 4291 went up in flames" on the night of 3 August 1918, as he had left Killingholme

Above: Another view of Felixstowe Naval Air station showing the large hangars, wooden slipways, flying boats on the hard-standing and overall size of the complex that was overseen by John Porte.

for his son George was born on that date, "but that night a wonderful American pilot went to see how much petrol was in the tank. He was unable to see the bottom, so he struck a match, with fatal results." If this happened to N4291 it was not fatal for the old boat, as it was finally written off in January 1919.[35]

The actions of the relative tiny flying boats against the giant Zeppelins were colourful but were not made public in order to keep the Germans from realising that their Zeppelins were now in danger. The pilots involved would have been feted by the public as the general population held the Zeppelins in particular odour as terror weapons or "Baby Killers." The other main tasks of the flying boats were hunting submarines and convoy escort, to be joined by long range reconnaissance missions as the capabilities of the boats were developed. J. L. Gordon considered that the "Curtiss Aeroplane Co. in the U.S.A. were the first to produce a flying boat which, powered by Rolls Royce motors, was capable of performing the many and varied duties required by the navy, with a moderate degree of safety and efficiency. It was the advent of this type of seaplane which altered the whole aspect of the work connected with naval co-operation."[36]

Despite the number of submarines sighted and bombed there is only one confirmed sinking of a submarine by a British aircraft without surface help for the whole period of the war. This action was fought on the morning of 22 September 1917, by H-12 serial 8695 operating from Dunkirk. This boat, crewed by Flt Sub-Lt N.A. Magor and Flt Sub-Lt C.E.S. Lusk[37], was to patrol near the German seaplane bases in Flanders and therefore had an escort of two Sopwith Pups. The Curtiss boat sighted and attacked the *UB32* that was spotted on the surface. The submarine emergency dived but two 230-lb bombs struck just aft of the conning tower and exploded. *UB32*, a Class UBII submarine, did not return to base.[38]

The flying boat had been sent to Dunkirk after seaplane operations were cancelled due to the operations of the Zeebrugge Friedrichshafen FF.49C and Brandenburg W.12 floatplanes. Magor and Lusk's H-12 arrived on 11 July. The seaplane pilots that were operating from Dunkirk now flew with escorting Sopwith Pup fighters. Capt C.L. Lambe, the OC Dunkirk, would come into criticism later in 1918 for not arranging for Dunkirk aircraft to rendezvous with flying boats sent out from the UK. (See Chapter 7). This would be another incident whereby the people at the Admiralty had no practical knowledge as to the capabilities of the aircraft in their charge. Aircraft "were mere pawns in the great game contested by the two fleets."[39]

The importance the Admiralty thought of aerial patrols in combating submarines is given by a Confidential and Immediate Memorandum of 5 July 1918, where in it was stated "a great naval effort is being made during the summer months to attack enemy submarines passing Northabout, and it had been hoped to obtain considerable assistance from air patrols. Their Lordships have been disappointed in this respect,

Above: Additional views of Felixstowe Naval Air station showing the large hangars, wooden slipways, flying boats on the hard-standing and overall size of the complex that was overseen by John Porte.

and trust that every endeavour will be made to ensure facilities for the maintenance of efficiency being provided before the summer is over."[40] This request was made in reply to a report on the Northern Patrol Seaplanes that revealed that in June Houton, Stenness and Catfirth stations had nine *Large America* boats but none were available for patrol. There were in addition four 240 Renault Short Seaplanes that were worn out and not much use for patrol work. The whole output of F.2A boats was reserved for Home waters, the F.3 going to the Mediterranean.

On 28 September 1917, Flt Sub Lt B.D. *Billiken* Hobbs, Flt Sub Lt R.F.L. Dickey and crew took off in H-12 serial 8676 at 0720 in answer to wireless intercepts. The British could pick up the German U-Boat traffic and figure out where the submarine should be located and then direct ships and aircraft to that position. Hobbs was looking for a submarine near the North Hinder light vessel. The North Hinder was sighted at 0805 when course was altered south. As a hostile wireless signal had been reported in a position 25 miles south of the North Hinder, the W/T operator listened in and at 0828 reported hostile signals being received by some vessel within ten miles. At 0834 an enemy sub was sighted in full buoyancy about one mile dead ahead. She was painted light grey colour and seemed to have a raised bow and also a mast and a gun. A man was observed forward by the gun. The seaplane increased speed to 80 knots at 600 feet and when about a quarter of a mile away fired two recognition signals which were not answered by the sub. Flying directly over the sub, the seaplane dropped one 230-lb bomb & then turned to repeat the attack, the sub at the same time firing one shell which burst 50 feet in front of the machine. The bomb was observed to make a direct hit on the tail of the sub, the explosion making a large rent in the deck. Whilst the seaplane was turning, a photograph was taken of the sub as she was under the port wing. At this moment several red flashes were observed on the water some distance in front of the seaplane and then through the mist three more enemy subs were seen heading SW in line abreast, and immediately behind them were three enemy destroyers. All six vessels were firing at the seaplane, but their shells exploded in front of the machine. Escorting the destroyers were two seaplanes which, however, did not attack the *Large America* owing to the barrage put up

Above: A Curtiss H-12 with bombs aboard on the hard-stand at Felixstowe. Flying boats required a hard surface for handling on land unlike their land sisters that could land most anywhere.

by the destroyer's fire. The *Large America* turned completely round and again passed over the sub which was now sinking by the stern with water up to the conning tower and nose out of the water. The second 230-lb bomb was released and exploded dead on 15 feet ahead of the bow. With the impact of this bomb the whole sub seemed to vibrate, and then sank immediately, leaving a large quantity of blackish oil, air bubbles and foreign matter. The crew were credited with sinking the *UC-6*, however, again post-war records showed that this was not the case.[41]

8676 delivered to Felixstowe on 15 March 1917, for erection. She had an impressive career and was credited with sinking *UC-1* with 8689 and N65 on 24 July 1917, and *UB-20* with 8662 on the 29th, and *UC-6* as recorded above, as well as two unsuccessful attacks. She sank when under tow by TBD *Meteor* on 27 December 1917.[42]

The arrival of the *Large America* boats had another benefit for the RNAS personnel who flew the floatplanes on patrols. Flying floatplanes was always dangerous for if they were forced down through any reason they were unable to stand up to the conditions usually prevailing in the North Sea. This was demonstrated on 24 May 1917, when Flt Sub Lt H.M. Morris and his wireless operator, AM2 G.O. Wright, left Westgate, a seaplane station on the East Coast, south of Felixstowe in their Short 827 floatplane serial 3072. After some hours patrolling over the North Sea, Morris turned for home. His engine began to miss then stopped altogether and he came down on the water. Unfortunately he had alighted in one of the British mine fields. It was a very big mine field, "starting from an east and west line a short distance south of the North Hinder and continued to a line running east just above the North Foreland." There were no ships in sight and, obviously, not much chance of a ship coming across them.

The sea increased in intensity so Morris dropped his bombs and let the petrol out of his tanks, lightening the seaplane. By four o'clock in the afternoon the wind had increased such that the machine was blown backwards placing the tail float in the waves. "The necessity of a tail float is the weak spot in the design of a float-seaplane, and the sea was attacking the flaw in the design." The crew took up station on the nose of each float in the hope that they would keep the tail float out of the waves. After about an hour the tail float gave way and inexorably the tail of the machine sank into the water and the machine capsized slowly backwards. The sea then proceeded to break up the machine so that the two were left with only a float to hang onto. The attachment points for the struts that supported the float gave the men a handhold. They were unable to climb onto the float as it was very unstable and so they were lying across it, half in and half out of the water. They clung to the float that night and the next day. The 26th saw a thick fog that eliminated any chance of a passing ship seeing them. By the 27th they suffered from thirst but the sea had calmed such that they could lie on the float. When the machine was wrecked they had suffered cuts and lacerations and these became swollen and

Above: Curtiss H-12 8654 with major damage to its starboard wing photographed at Tresco. Delivered in December 1916 the machine was at Tresco by late February. Two months later it was damaged in attacking a U-Boat that fought back and was unable to drop its bombs. On 21 July 1917 it was struck by 8686 taking off and damaged. Deleted in July 1918.

inflamed. On the 29th, after spending five nights on the float, they were very weak. At noon the fog came down. They were so thirsty they began to take sips of sea water.

At Felixstowe Hallam ordered two boats run out even though it did not look promising. At 12.17 the boats were put into the water and took off. The fog became so thick that one boat turned back. The other, flown by Flt Sub-Lt G.R. Hodgson and Flt Sub Lt J.L. Gordon, decided to press on. Eventually they also had to turn for home so thick had the fog become. At an altitude of 1,200 feet they were about 12 miles from the North Hinder when they spotted something on the water. Spiralling down to 600 feet they saw two men on a float. A strong wind was blowing and a heavy sea was running. It was obvious that the two men were in dire need of assistance.[43]

Hodgson later recalled:

The first important problem we encountered was landing a boat in the open sea. A gentle sea swell was running and a very light breeze blowing. The latter probably forced us to contact the water at 50 to 60 knots.

To say we got a bad shaking up is to put it mildly but so far as we then knew no damage had been done.

The first attempt to take the men failed. The boat was taxied up to the float and the engines stopped at the very last minute, however the wind blew them away from the pair on the float. A second attempt was made with two of the flying boat's crew standing on the fins each side of the bow. Waves were washing up to the waist of these men but they managed to seize Morris and Wright and drag them up to the drift wires that connected from the wings to the bow of the hull.

Inside of five minutes Morris and Wright were in our boat.

Then began the time of decision: Should we try and take off or make a taxi run to the East Coast. We decided on the taxi run, but a few minutes after starting our engines Anderson told us that on landing we had broken a hole in the bottom of the boat and we were taking in sea water. We realized of course we were in a tough spot and our only hope of getting out of it was to take off. Our attempt was not a success and we were compelled to abandon it. One of the things we did was to break the tail plane.

We then returned the taxi run hoping for the best. We got to the shipping channel and there we had the good luck to meet a small cargo vessel which picked up all of us, put a line on our boat and took us back to Felixstowe.[44]

Hodgson had fed the two rescued men brandy from an eye dropper until they were able to take some warm cocoa from a thermos flask. The *Orient* out of Leith picked up their distress signals and gave them their initial tow. Their bilge pump had failed and they had to bail by hand, and also pump the fuel to the engines by hand. The sea was picked up by the airscrews and thrown over them, coating them with salt. The tow was

Above: This H-12 at Tresco is being dismantled/erected on the beach. The upper surfaces of the wings and tail surfaces are a dark (green) colour while the hull appears to be the usual two colour Curtiss standard.

Below: An H-12 at Tresco.

Above & Facing Page: Close-up views of the crew in an H-12. Note the starboard pane has been removed from the windscreen. The wind driven fuel pumps can be seen behind the main cockpit.

later transferred to *HMS Maratina*, while Morris and Wright were transferred to *HMS White Lilac* and rushed back to port for medical attention.[45]

James *John* Lindsay Gordon and George Ritchie Hodgson were two Canadian cousins who had been boys together and had come to the UK together, learned to fly together, and when they came to the Felixstowe War Flight asked to be able to fly together. Known as "The Heavenly Twins" they flew together for some time alternating First and Second Pilot roles, but both were good boat pilots and these were in short supply and so they were each given a boat. At first they resented being separated but finally saw the necessity of the move. They had their names bracketed for Duty Pilots and for leave and usually managed to fly their boats in company. On this day they had managed to fly together once again.

> *Hodgson was had been a champion swimmer. He was a stout fellow in more ways than one, and had been built for big boat work. Gordon was a long-faced, serious lad, not over strong physically, but with tremendous determination and force, and was a careful flying-boat husband. Both men were great grumblers, but also great workers.*[46]

Hodgson had won Olympic gold at the 1912 Stockholm Olympic meet in the 400 metre and 1,500 metre swim events. The two Canadians received the British Board of Trade's Silver Sea Gallantry Medal for the rescue. Both survived the war, Hodgson with the AFC and Gordon with the DFC. Gordon was to remain in the Canadian Air Force becoming Director from 1922 to 1924, and RCAF Senior Air Officer in 1932. He passed away in November 1939 at the age of 48 years. Within two months Morris and Wright were returned to duty.

Sir Austin Robinson wrote to Hodgson in 1981 that with respect to their rescue of Morris and Wright,

> *I do not think that you can ever have appreciated yourself what your rescue of them* (Morris & Wright) *did for all the rest of us who were flying seaplanes over rough seas. I do not have to remind you that in those days it was the sea that was the enemy*

and the morale of a station enormously depends on whether pilots who have been lost through forced landings from which they were not rescued. The feeling that in such an emergency your friends would come to your rescue made a tremendous difference to morale.

H-12

Preceding Pages, Left Page and Top Right: On the early H-12 boats the equipment seems to have been added wherever it could be fitted. There are two pillar mountings for the front Lewis gun. The ladder gives access to the walkway over the fin. The stencil on the hull above the walkway reads "DO NOTE WALK ON FIN." The cradle is marked 86X3 or 86X8 which establishes that this is a machine from the first batch of H-12 boats delivered. This machine has the serial marked in a dark rectangle.

Preceding Pages, Bottom Right: The handed, inward-turning airscrews are well illustrated here. These blades do not appear to have brass tips or fabric coverings. Note the bomb racks under the lower wings and bomb sight on the nose.

Above: The early Curtiss airscrews caused problems when they threw blades as shown here.

Below: An early H-12 moored out.

Above: This H-12 has a gun ring fitted with twin Lewis guns but the additional Lewis at the cockpit is still carried. Note the bombsight on the bow and the height above the hull that the wind driven fuel pumps props were located.

Below: Twin Lewis guns and a large bulky camera. It is not known if the camera was attached to the gun ring. Losing it in flight would cause it to be thrown into the airscrew with disastrous results.

Above: Bombed up early H-12. Note the white serial on black rectangle to hull.

Below & Bottom: The Curtiss wings were employed for the Felixstowe boats and were identical until the introduction of balanced ailerons on the F Boats as shown here. The Felixstowe boat's airscrews rotated in the same direction.

Above: Another early H-12 at moorings. These boats appear to have had the dark rectangle applied to the rear hull for the serial application. On this machine the "No Step" legend is to the right of the walkway over the fin. Note bomb under lower wing.

Below: Working on an H-12. Note that the rear of the hull has to have a prop to stabilise the machine. This machine has the early pillar mounts. The engines are running and it is armed with bombs and it is possible that it is getting ready for launching, the engines being warmed up before the boats were launched.

Above & Facing Page: Views inside the H-12's hull. The lack of dual control is apparent. The second pilot/front gun layer's passage was on the left. In the Felixstowe F.2A the second pilot's seat could fold out of the way to allow access. The fuel tanks, located between the wings, separated the crew from each other.

I had only one forced landing myself in an H12 and I was certainly no more successful than you were (in getting off again). *In my case, I got the H12 down in a rough sea without breaking the hull. But we could not get off again and within a couple of hours one wingtip float was broken and the fabric carried away from both lower wings. So any further attempt to take off was out of the question. With two members of the crew sitting out on the wing that still had a wing tip float we taxied safely ashore on to a sandy beach somewhere south of Bridlington.*

The design of the H12 was surely at fault. It had a wonderfully strong forebody and the woefully weak monocoque tail which always broke just aft of the step. Any forced landing in an H12 was likely to end in fracture there. Equally Morris would never have been in the predicament he was if it had not been for the bad design of all Shorts. They always stove in the tail float and turned topsy turvey in a forced landing in a rough sea.[47]

During October 1917 the weather was so bad that little flying was done from Great Yarmouth. 21 year old Flt Sub-Lt Peter George Shepherd with his observer, 20 year old L Mec Walter Fairnie, were lost in Short 320 floatplane N1360 on the 28th. They did not return from patrol and were never seen again. The next day Bob Leckie, Flt Sub-Lt Bolton, CPO Whatling and AM Walker took off in H-12 serial 8660[48] to search for the missing aircraft. They came across an enemy submarine when Whatling sighted a conning tower break the surface. Leckie turned to attack but the submarine had dived. Spotting the periscope which popped up for a short time they attacked, Bolton releasing two 100-lb bombs with 2½ second delay fuses, but no visible results were obtained.

8660 had a distinguished career attacking submarines on two occasions and the Zeppelin L46. Wrecked on 6 November 1917, she was rebuilt with a F.2A hull and returned to service in January 1918. She met her end on 30 May 1918, when, with a mixed British and USN crew, she was attacked by enemy seaplanes. On that day F.2A serial N4295 and 8660 left for a long patrol. Ens George Thomas Roe was 2nd pilot to Capt Charles Leslie Young, DSC, in 8660.[49] They were forced down with engine trouble and but were able to signal the other boat by Aldis lamp that the fault was repairable. The F.2A circled while repairs were carried out. After about 50 minutes two German seaplanes appeared and joined combat. The front guns of the flying boat jammed after a few rounds. The German seaplanes turned away and headed for Borkum for reinforcements. The F.2A set off in pursuit but soon realised that it was being led away from its charge. Returning to the area he had left 8660, the boat was nowhere to be seen and so N4295 returned home to Great Yarmouth. A boat was sent

out to look for the missing Curtiss. It had not long left when a pigeon was received with the message that 8660 was on the water and under attack.

8660's crew had repaired the fault and taken to the air again. After about 45 minutes, the boat was again forced to alight. She was on the water when five German seaplanes attacked. The British boat fought back, the attack continuing until the Germans saw three men jump overboard and swim away from the flying boat. A German seaplane immediately landed and the gunner boarded 8660. He found the pilot dead and the mechanic wounded by splinters. Two of the swimmers were rescued, the third disappearing. The boat was set on fire and the Germans returned with their prisoners to Borkum. The dead pilot was Young, and the mechanic who drowned was AM1 Henry Francis Chase. Roe, Cpl F Grant and Pte JN Money were taken POW.

Roe was to spend the rest of the war as a POW in Camp Lanschut, Germany. He was awarded the Navy Cross for his part in the action. Roe was killed on 28 May 1921, when he crashed in Loening M-81 monoplane A-5762 at NAS North Island, San Diego. The aircraft was on a test flight at 3,000 feet when it was observed to enter into a series of involuntary spins. Roe recovered from all but the last, striking the beach at terrific speed. His passenger, CMM J.P. Dudley, was seriously injured but recovered.

Upon the report of a *Large America* boat exploding in the air in the Scillies, Yarmouth reported that there was a noticeable tendency for petrol vapour to mix with air and accumulate in the bottom and tail of the hull and unless the hull was freely ventilated there was the danger of a highly explosive mixture accumulating there. It was noted that Felixstowe was looking at the problem.[50] No modifications to prevent a repeat of this incident have been discovered to date.

Danger was not only experienced in the air. On the morning of 4 July 1917, some 12 hostile aircraft raided Felixstowe Station at 7.25 B.S.T. One bomb fell on the beach close to No.2 Slipway setting fire to the Curtiss *Large America* 8679, and causing several casualties amongst the ratings. Another bomb fell in the water near the slipway by the old Station, causing no material damage but more casualties. A third bomb fell amongst the new huts being erected across the road injuring two of the contractor's men.

Above: The Felixstowe F.2A pilot's instrument panel. Via AHT AL0655-009B.

In all four men were killed; 15 severely wounded, and four slightly wounded. In addition two civilians were killed and two wounded. Curtiss H-12 No. 8679 was completely destroyed and 8682 suffered considerable shrapnel damage. The top part of No.3 Slipway was badly charred for about 12 yards.[51]

The raid was carried out by Gotha bombers under the command of Rudolf Kleine, 18 of the bombers actually made the attack out of the 25 that started out. The Home Defence organisation was laggard in its response and only one aircraft attacked the straggling formation on its way back across the Channel. This was a RAF powered D.H.4, serial A7436, on a test flight from the Testing Squadron, Martlesham Heath. When he came across the raiders the pilot, John Palethrope, accompanied by AM1 James Oliver Jessop, attacked the middle of the formation. Coming under intense fire his gunner was hit in the heart and his front gun jambed. Paleathrope returned to land and took off again with another gunner but did not reengage the Gothas. A total of 17 were killed and 30 injured in the raid.[52]

Enter the F.2A

The Felixstowe F.2A was the result of John Porte's experiments on flying boat hulls as detailed in Chapter 2. It was to become the most famous type of British flying boat of the war due to its combats with Zeppelins and German marine aircraft as well as its use in long-range reconnaissance patrols and anti-submarine work. The first Production F.2A delivered was N4280, a Saunders built example. This machine was delivered to Felixstowe during the week ending 17 November 1917.

The F.2A worked with the H-12 and H-12 Converts from those stations where conflict with enemy aircraft could be expected. The F.2A overcame many of the problems that had arisen with the Curtiss boats and proved so successful that it was to be built in the USA with Liberty engines as the Curtiss H-16. The circumstances behind the decision to build the F.2A in the US are unknown. Did the British supply Curtiss with plans that enabled that Company to build the H-16 before the US NAF entered into the discussion? The British ordered the H-16 in June 1917. British H-16 boats entered service around

Above: The fuel tanks, located between the wings, separated the crew from each other. Via AHT AL065-009C.

April 1918. The fact that the NAF redrew the F.2A plans for manufacture in the US points to their involvement at an early stage. This question requires more research.

The H-16 was used at Dundee from where N4891 made a 10¼ hour flight on 27 August 1918. The flight was the longest that was recorded by any H-16 or F.2A but ended badly. The boat crashed while landing in the dark on the River Tay writing the boat off but with no injuries to the crew. Dundee was being reinforced in the autumn of 1918 as it was suspected that the German High Seas Fleet might make a last sortie. The flying boats were ready to assist the British fleet in whatever arose. By the end of December Dundee had 12 flying boats, four H-16, five F.2A and three F.3 boats. On Armistice Day Sir A. Robinson was shipping N4484 to Dundee.[53] On 3 June N4066 from Dundee, with the crew of Capt W.R. Kenny, Capt N.H. Woodhead, AM1 C.G. Allred and AAM1E .P. Denton, dropped their two 230-lb bombs on a submarine that had submerged on their approach. Dundee did not get the acknowledgement that Felixstowe and Yarmouth received but it carried out its job with a variety of aircraft.

The USN had considerable difficulty in bringing the H-16 into service and at Killingholme H-12, H-12 Convert, F.2A and British H-16 boats were utilised by that Station until their H-16 boats entered service. Short floatplanes were also flown operationally from Killingholme. Norman Thompson N.T.4 and Porte Baby flying boats were also recorded by NAS Killingholme, and if flown would have been only for training flights.

The following precises of official reports concerning N4287 from Killingholme Station when it was in the process of being handed over to the USN, gives an indication of the work done by the Felixstowe boats.[54] F.2A N4287 was from a batch of 30 flying boats built by S.E. Saunders Ltd at and built at Cowes, and delivered to Killingholme via Felixstowe on 20 March 1918.

17.06.18. 1st Pilot Lt Cleeve, RAF; 2nd Pilot Ens Allen, USNRF. Sighted suspicious oil patch 15 miles due East of South Creek at 0755. Dropped two bombs one of did not explode. No results observed.

26.06.18. Large America 4287. Ens J.J. Schieffelin, USNRF

Above: Felixstowe F.2A interior looking forward.

(pilot).[55] Whilst on patrol sighted the conning tower of sub three miles distant in Lat 53° 33Z' N, Long 0° 59' E. The sub submerged rapidly while the seaplane was still 1½ miles away and started to move slowly westward. Two 230-lb bombs were dropped on position sub submerged which burst 150 and 200 yards from the hull. Nothing further observed and seaplane returned to base. Only the conning tower, that was painted black, was seen.

28.06.18. 1st Pilot Lt (jg) F.R.V. Lynch, USNRF; 2nd Pilot Ens Grosvenor, USNRF. Whilst on patrol at 1120 sighted a submarine awash which immediately submerged. Two bombs were dropped from an altitude of 400 feet which fell on either side of the course of the submarine had apparently taken. The position of the sub was sent to base and after circling the position for 15 minutes and observing nothing further the seaplane returned to base.

13.07.18. Lt Lawrence (pilot), Ens G.S. Hodges (pilot). Whilst on patrol at 0714 sighted the wake of a submarine. One bomb was dropped after passing over the submarine twice, bubbles rose from the submarine. A second bomb was dropped and an oil patch was observed on the surface in the position where the bomb was dropped. On being informed by the seaplane, a destroyer came to the position of the submarine and dropped 6 depth charges. The seaplane left the destroyer patrolling the vicinity and returned to base.

10.08.18. 1st Pilot Capt Pattinson, RAF; 2nd Pilot Lt (jg) Lawrence, USNRF. Position 54.12N, 04.10 E, sighted Zeppelin to the North at a distance of 20 miles. Zeppelin first seen steering NE at about an altitude of 8,000', the seaplane at 6,000'. Altered course. At 1725 broke off chase, Zeppelin at 9,075'. Distance from Zeppelin 2 to 3 miles. Distance covered 504 miles.

N4287 was turned over to the USN in July 1918, and was written off in Week Ending 5 September 1918.

The F.2A boats were more capable of taking care of themselves than the H-12. On 18 March 1918, Great Yarmouth F.2A N4512 left at 10.30AM with Flt Cmdrs Bob Leckie and G.E. Jerry Livock[56] as pilots, and Flt Lt Daddy Brenton replacing the usual gun-layer, AM West, with PO Thorpe and AM Chapman completing the crew. When some 10 miles from Terschelling they sighted two hostile seaplanes and they promptly gave chase. Livock dropped their bombs

Above: An USN Curtiss H-16 of Killingholme station. This machine bears the Station code "K"; the last numeral of the individual aircraft number cannot be positively identified. The machine is Navy Grey overall with US insignia. Note that the top of the radiators appear to be blanked off and the boat is on the cradle for K-11.

to lighten the boat. In the resulting running battle, during which they fired nearly all the ammunition for the forward guns which jammed every few seconds. Hits were observed on one HA but did not hit anything vital as it kept on its course. After about 15 minutes Borkum Island was in sight so it was time to turn for home before reinforcements arrived from its air station. "There was no manoeuvring or anything like that – the two machines just flew along side by side pumping lead into each other as hard as they could." The enemy observer was an excellent shot. Their boat had been hit in several places. One fuel tank was holed and the port oil tank punctured. After about a half hour they gave up the chase.

At 12.45 three enemy seaplanes were seen and course was altered to avoid them given the condition of the boat. The Germans attacked one proving very accurate as they hit the fuel tanks again. The boat had dropped to within five feet of the sea with a German "in station directly behind us." This machine continued to fire short bursts hitting the tail. At 1.10PM the enemy gave up and headed for home.

The German's method of attack was to get directly behind the boat on the same level. "This is practically the only blind spot in the boat, and, unless by yawing the machine about, we were unable to bring our guns to bear upon him." This was still a vast improvement over the H-12 that was completely blind from below and behind. The boat arrived back at base with only 50 gallons of fuel left, another 50 gallons was estimated to have flowed out of the tanks into the boat's bilge. This was

Above: N4067 was one of the few Curtiss built H-16 that saw actual service in British hands. Delivered to Felixstowe in April 1918, this boat was sent to Killingholme in June where it was allocated to the USN. It crashed on the way to Dundee on 20 October 1918. It is assumed that this is the incident in the photograph. The first pilot and engineer are thought to have died in the accident; unfortunately, the names of the crew have not been located to date. (Ens Raymond L Atwood collection/Emil Buehler Library)

the first encounter of the Great Yarmouth boats with German fighting seaplanes.[57]

Danger did not always come from the enemy. Flt Lt Daniel Fairman Ellis took Ens H.M. Wilcox[58] up in an H-12 8669 on 9 March 1918, apparently as a check flight for Wilcox. There was an accident with the airscrew that according to Wilcox led to the loss of both Ellis's legs. Wilcox remembered that "When we were in the water, the Canadian pilot complained 'Something is wrong with my legs.' I managed to remove him from the cabin of the plane and carry him piggy back out on the wing." A tug came and rescued them and Wilcox accompanied the injured pilot to hospital.[59]

Above: The Friedrichshafen FF33L was a smaller, faster variant of the FF33 series and 40 of the 125 built were configured with a fixed pilot's gun in addition to the flexible observer's gun as shown here. The limited top speed of the FF33L, 86 mph, made intercepting the *Large America* flying boats very difficult. This is the prototype.

Above: The arrival of the Brandenburg W.12 at North Sea flying stations posed a much greater threat. More maneuverable than the FF33L, the W.12 also possessed a higher maximum speed of 99 mph, making it much easier to intercept the big flying boats. The FF33L and W.12 are likely the types encountered by F.2A N4512 on 18 March 1918.

On 25 April 1918, two boats from Felixstowe were attacked by the redoubtable *Oberleutnant d R* Christiansen's Brandenburg monoplanes from Zeebrugge.[60] Curtiss H-12 Convert 8677 was accompanied by F2.A N4284. They left Felixstowe at 3.55PM and at 4:33 encountered seven hostile machines and attacked them at close range. The combat report from the crew of N4284, Capt Young and Capt Tees with Engineer Edwards and W/T Operator Nicol, described the attack as follows:

Visibility about 8 miles, hazy in places. Submarine patrol in company with machine 8677. At 4.53 when flying close to leading machine (8677) and slightly astern on port quarter, we sighted enemy seaplanes on our port bow flying towards us. Our machines were approximately at 800 ft. and both machines simultaneously turned left and dived to attack them at very close range. 8677, passed over us on the turn and we completed the turn to keep formation with our leader, when we observed that he had dived to within 100 Ft of the water, and was a short distance ahead steering West at full speed. We opened engines full out, and gradually drew abreast of 8677, on their Starboard side. As our machine was the faster, we throttled down to regulate our speed to that of 8677 and we kept abreast, and very close together, fighting the seven attacking machines with our rear guns.

After flying very low in this formation at about 75 Knots for about 4 minutes, 8677 dropped a little astern, so we throttled down to 70 knots to enable her to draw abreast. As she did not catch us up after a minute, we still further throttled down and also reduced out (sic) *ground speed by climbing our machine.*

At 5.32, 8677 which was about 90 to 100 yards on our port quarter, was observed to turn to port and immediately burst into flames and completely disappear under water.

The enemy continued to attack us and as they were faster than out (sic) *Machines, we opened full out, dropped our bombs and nosed down to 50 ft.*

At 5.8 when by Outer Gabbard Bouy, the enemy turned East and we returned to Felixstowe. No signals were received from, or sent to the leading Machine (8677) after sighting the enemy.

The Enemy machines were fast single and two seaters, we were unable to ascertain the total number of enemy machines until they were all attacking us in the rear, but we do not think we reduced their number when we made our initial diving attack, although the machine we concentrated our forward gun fire at very close range, banked steely away from us and was lost to view as we turned after the leading machine.

When attacking the enemy our front gun jambed after firing a few rounds from the second ammunition tray.

The port rear gun also jambed when we were being hotly attacked and this jamb took some time to clear as it was necessary to remove the Scarff sight to do so. There were also 3 other jambs in the port rear guns.

As the enemy usually attacked from a position directly behind our tail the top rear gun was practically useless.[61]

The Germans also left an account of this action:
After fire had been opened on the flying boat from aft by the fixed machine-gun, and the machine-gunner in the stern of the flying boat killed, Christiansen flew parallel with the boat, and

Above: Line up of six Felixstowe F.2A flying boats on the slipway at Felixstowe.

Below: Loading a bomb on to an F.3 *Large America* flying boat. The crude method utilised here may be compared with that displayed in other photographs. Note the open hatch near the wing leading edge.

Above: N4297 in one of the "standard" colour schemes applied to the F.2A boats.

*his observer (*Vizeflugmeister *Wladicka) opened fire at the oil tank at the rear of the port engine, setting the tank alight. The pilot of the flying-boat then tried to alight, but as his machine was only 10 feet from the water, he could not turn her into the wind, so that she crashed on alighting and burst into flames. Of the crew, 3 men were seen swimming about, but the sea was too bad to allow of our machines alighting in the hope of saving them.*[62]

Capt N.A. Magor, Ens Steven Potter, USN, and their crew, Cpl Reginald Arthur Lucas and AM John Gardner Strathearn, all perished.[63]

Potter was Naval Aviator No. 130, and a member of the Second Yale Unit. He trained with the Unit on Curtiss F-Boats before gaining his wings on 29 October 1917. He attended Moutchic in France for advanced training and was credited as the first aviator to bring down an enemy aircraft, a German seaplane on 19 March 1918, as related above. Potter wrote to his family just nine days before his death: "If you receive this, you will know I have done my duty to the best of my ability. Be sure I am wonderfully glad that I could give up my life so usefully."[64]

A note to the file reads "There seems to be something wrong with either the ammunition or the guns." There was to be Court of Enquiry into the loss of 8677, however a copy of this has not been located as yet.

Not all patrols led to engagement with the enemy submarines or aircraft but they were still dangers always present. On 2 October 1918, six F.2A boats left Felixstowe at 07:30 to carry out a long reconnaissance patrol to the Hook of Holland up to the Coast to Texel and return. When about

7–8 miles S.W. of Yumiden F.2.A 4551 was forced to land on account of engine trouble. The other machines stood by and saw the engines being started up again and the machine leave the water O.K. It was forced to land again soon after, and the other machines after circling were obliged to leave owing to the long distance they had to return and the shortage of petrol. F.2A 4551 was drifting towards the Dutch coast when last seen. The five machines returned safely to base with the exception of F.2A 4301, which had to land owing to running out of petrol, about 20 miles East of Orfordness. She was towed back safely to harbour by a M.L. arriving at 1640. The other four landed at their Base at 1240.

Details of the flying boats that took part in this patrol are given below:

N4297: Pilots Capt Webster and 2/Lt MacSwiney; Observer Lt Cotton; W/T Operator Norris, and Engineer Kirby.

N4545: Pilots Lt Boulding and Lt Harrison; Observer Lt Roper; W/T Operator Searle, and Engineer Clarke.

N4537: Pilots Lt Freeman and Lt Rogers; Observer Lt Keddy; W/T operator Coleman, and Engineer Chrismas

Above: F.2A N4510 at Grain on 15 February 1918. Note the square sliding door to the hull side windows.

N4301: Pilots Lt Hunt and Lt Browne; Observer Lt Frost; W/T Operator Putman, and Engineer Evans. Alighted due to running out of petrol. Fired Verys lights and picked up by M.L.20 that towed the boat to the Orfordness Light where the tow was turned over to M.L.58 which towed the boat back to base.

N4551 was salved by the Dutch and given the Netherlands Naval Aviation (*Marine-Luchtvaartdienst* – MLD) serial L.2. The crew of 2/Lt T.N. Enwright, 2/Lt W. Pendleton, 1AM Henry Lawson Curtis and AM William Arthur Mitchel, was interned.

The same day Felixstowe sent out a second patrol of three F.2A boats at 0830 to conduct a search for a submarine.

N4539: Pilots Lt Land and Lt Shield; Observer Lt Moir; W/T Operator Beeston, and Engineer Cpl Pretty

N4304: Pilots Capt Bailey and Capt Nemus; Observer 2/Lt Irvine; W/T Operator Miles, and Engineer Robinson

N4530: Reported no engineer aboard was ordered to return to Felixstowe.

N4530 was forced to land off Southwold on account of engine trouble and was eventually towed to safety. The remaining two carried out the patrol and returned to harbour with nothing to report.[65]

"Nothing to report" is seen more often than combat activities in the daily patrol reports of the flying boats from all stations. Great Yarmouth, Felixstowe and Killingholme were in the areas of most activity. The bases in the north and Ireland flew their anti-submarine and convoy protection patrols with very little to end the monotony of these long, lonely patrols.

On 30 May a Yarmouth boat, 8660, had been forced down with engine trouble and set alight on the water by the Germans as recorded above. It was thought that the Germans were faking Zeppelin communication wireless traffic to lure the British flying boats into the area where the German floatplanes then sprang a trap. On 4 June 1918, Yarmouth sent two F.2A boats to meet up with three Felixstowe boats. The Felixstowe boats were:

N4302 under Capt A.T. Barker (1st pilot); Lt .Vernon F.A. Galvayne (2nd pilot); Ens Kenneth Burton Keyes, Pte Hopkins (W/T operator); Pte Reid.

N4533 under Capt R.F.L. Dickey; Capt R.J. Paul; Lt A.G. Hodgson; 2AM E.P.C. Burton and AC H. Russell (W/T operator).

H-12 Convert serial 8689 under LT M.J.R. Duff-Fyfe; Flt Sub Lt J.R. Pattison; Ens J.A. Eaton, USN; 1AM E.J. Strewthers[66] and W/T Sgt A.J. Browne (W/T operator).

While the boats from Yarmouth were:

N4295 under Capt R Leckie with Capt Bolton, Maj Haggerston, Pte Deeley (W/T operator) and Pte Chapman.

N4298 Capt John Hodson with Capt Mossop, Pte Taverner, Corp Beaumont, Pve Raymond (W/T operator).[67]

The five *Large America* boats set out to engage the Germans "in the appropriate manner." Now the British were seeking revenge and they flew into what was possibly the largest action

Above: By November 1918 N4297 sported a dazzle colour scheme that incorporated the wingtip floats.

of the war between the British flying boats and their German floatplane opponents. Ens Keyes, USNRF, was the front gun-layer of N4302 and described the action as follows:-

Our three machines from Felixstowe rose from the water at noon, circled into patrol formation, and proceeded along the coast to Yarmouth. Here we were joined by two more planes. At one o'clock the squadron turned east. We sighted nothing until half past two, when the Haaks Light Vessel slowly rose on the horizon, and then a fleet of more than a hundred Dutch fishing smacks. Soon we perceived the Dutch coast and followed the sandy beaches of Texolt and Vlieland Islands until we came to Terschelling. We could distinguish houses, and make out breakers rolling up the sandy beaches.

At Terschelling we veered west, but we soon had to turn back because one machine (Watson's) had come down on the water with a broken petrol pipe. We circled it, and fifteen minutes later sighted five German planes steering west, which would soon bring them upon us.

Lieut. Galvayne was seated near the wheel. His duty was to kneel with eyes above the cowl, and direct the pilot. I was in the front cockpit, with one gun and four hundred rounds of ammunition. In the stern cockpit the engineer and wireless ratings were to handle three guns.

We took battle formation and went to meet the enemy machines. But when almost within range, they turned and ran away from us. At once we gave chase, but soon found that they were too fast. Our plane was nearest the Germans, so I had the satisfaction of trying out my gun for several rounds. It was impossible to tell whether I registered any hits or not.

We had chased these planes to keep them away from the machine on the water, which otherwise would have been shot to pieces. Finding now that they could keep out of range, we turned back, and again circled the disabled plane.

Soon the enemy once more came close, and we gave chase a second time. But instead of five machines, as before, there were only four. One small scout was sighted flying in the direction of Borkum. For the third time we went off in pursuit.

Suddenly we discovered that a large number of hostile planes were steering towards us, not in the air with the four planes, but very close to the water. Ten machines were in this group, but they were joined in a few minutes by five more. The scouts were painted black, the two-seaters green, and seemed very hard to pick up.

We swung into battle formation and aimed for the middle of the fleet. When we were nearly within range, four planes on the port side, and five on the starboard side, rose to our level of 1,500 feet. Two planes passed directly beneath us, shooting upward. Firing was incessant from the beginning, and the air seemed blue with tracer smoke. The Germans used explosive bullets. I gave most of my time to the four planes on our port side, because they were exactly at our level, and within good range – about 200 yards.

Once I looked round and noticed that Lieut. Galvayne was in

Above: A *Large America* on the hardstand area outside canvas hangars.

a stooping position, with his head and one arm on his seat, the other arm hanging down, as if reaching for something. I had seen him in this posture earlier in the day, so thought nothing of it. All this I noticed in the reaction of a second, for I had to continue firing. A few minutes later I turned around once more, and found with a shock that Lieut. Galvayne was in the same position. It was then that the first inkling of the truth dawned on me. By bending lower, I discovered that his head was lying in a pool of blood.

From this time on I have no clear idea of just what our manoeuvring was. Evidently we put up a running fight, steering east, then circling. Suddenly I found that our machine had been cut off from the formation, and we were surrounded by seven enemy seaplanes. We were steering almost south-west. We fought for ten miles or so, until we drove the seven Germans off. One of them was driven down out of control and made a very poor landing. Another was badly hit, side-slipped, and crashed in flames from a height of 2,000 feet. All were severely punished.

During the last few minutes of the fight our engine had been popping too frequently, and soon the engineer came forward to say that the port engine petrol pipe had been broken. By this time I had laid out Lieut.. Galvayne in the wireless cockpit, cleaned the second pilot's seat, and taken it myself.

The engagement had lasted about half an hour. We descended to the water at 4.45PM ten miles north-west of Vlieland. There I loosened Lieut. Galvayne's clothing, made his position easier, and felt for his heart, which I was sure was beating feebly. Then we rose to 1,500 feet, and sighted two Yarmouth planes. We picked them up, swung into formation, and laid our course for Yarmouth. At 7.10 we sighted land, and twenty minutes after were resting in front of the Yarmouth slipway. We at once summoned medical aid, but found that nothing could be done for Lieut. Galvayne. A shot had gone through his head, striking the mouth and coming out behind an ear, tearing a two inch gash.

Our boat was riddled. A number of shots had also torn the top between the front cockpit and the beginning of the cowl. The duration of the flight was seven hours and ten minutes.[68]

Capt A.T. Baker escaped unscathed from the combat despite sitting next to the 20 year old Lt Vernon F.A. Galvayne.

N4533 was the flying boat forced to alight with engine trouble. Ordered by the patrol commander to taxi to Dutch waters and destroy the boat on nearing land, the patrol was escorting the downed machine towards Dutch territory when the Germans came on the scene. The ordeal was not over for the crew of N4533 as it was now shot up while on the water some 200 yards from shore and set on fire by Christiansen in Brandenburg W.19 Marine Number 2239. The crew escaped and were interned. The Dutch purchased one of the Rolls-Royce engines salvaged from the burnt out remains of N4533. Curtiss H-12 Convert 8689 was so badly shot up that they were forced to alight in Dutch waters. Eaton reported that his machine and oil tanks were punctured, his fuel leads severed, and his glasses shot away. When the fight was over and they had started on the homeward journey the motors gave trouble and with the wind coming from off-shore and at 500 feet altitude, they were forced to alight in Dutch territorial waters. The boat was beached on Vlieland Island and the crew arrested by Dutch infantry before repairs could be effected. The machine and crew

Above: A *Large America* overflies a British submarine. In addition to hunting Zeppelins, anti-submarine, and convoy patrols, the *Large America* boats performed reconnaissance missions to determine the movements of German ships and the location of mine fields.

were interned.[69] 8689 was taken on charge by the Netherlands Naval Air Service (*Marine-Luchtvaartdienst* - MLD) as L.1.

Leckie reported after the action that it was "obvious that our greatest foes are not the enemy but our own petrol pipes."[70]

On 5 June 1918, another combat saw an instance of the chivalry that was to be displayed at times by the marine airmen. On this date Lt Col E.D.M. Robertson, Felixstowe's CO, was acting as 2nd pilot to Flt Sub Lt J.O. *Tiny* Galpin in H-12B N4345 and they found themselves in a fight with five hostile floatplanes. Forced to attempt to alight with an engine shot out the boat was wrecked on striking the water. The crew managed to climb out and climbed onto the overturned hull. One of the attacking Brandenburg floatplanes landed and taxied close to them. The German pilot asked if they preferred to wait in the hope that they would be picked up or would they like to surrender and be flown to Zeebrugge. Robertson politely declined the offer and said that they would take their chances of being found. The German then took a photograph and took off. Roberston and his crew were picked up the next day.[71]

According to Air Commodore Charles R. Samson, Sopwith Camels accompanied the *Large America* boats on anti-Zeppelin patrols. The boats could not reach the heights that the Zeppelins now operated at and hence the need for a fighter that could intercept the airships. The F.2A would patrol for miles into German territory accompanied by two or three Sopwith Camel fighters, the flying boat acting as navigator, a lifeboat if a Camel came down in the sea, and as bait to draw the German seaplanes out. "We hoped all they would see would be the F.2A and miss the little Camels which flew at a much higher altitude." Samson, despite his administrative duties, made several of these patrols and despite his assertion that the Camels "would have made short work of the German seaplanes."[72] an incident on 18 July 1918, showed otherwise. On this date two Short floatplanes were escorted by two Camels, the redoubtable Christainsen and his men shot down both Shorts and sent the Camels back to Manston shot full of holes.

The Felixstowe F.3

Confined to those areas where aerial opposition was unlikely, the F.3 did the mundane task of endlessly searching for submarines. These patrols could see some excitement as described below.

N4238 was one of a batch (N4230–N4279) built by Dick Kerr Ltd. On 17 May 1918, Capt Fitz-Randolph (pilot), Lt Bell (observer), and crew from Great Yarmouth, were on patrol with F.2A N4295 when they heard loud enemy W/T transmissions, but after climbing to 10,000' in good visibility, no enemy craft were sighted. They altered course and a U-Boat, apparently fully blown heading in an easterly direction, was sighted in position 04.30 W, 54.38 N, five miles distant on the port bow of the seaplane. 4295 descended to 1,800 feet and turned to bring the submarine on her starboard bow. One

Above: This late model F.2A without the cockpit canopy bore the name *Old Blackeye I*. This machine has been identified as N4082 that was allocated to the USN at Killingholme. Sir A. Robinson identified it as H-12 Convert 8688 in a photograph in *Cross & Cockade Great Britain*, Vol.11 No.1 1980. He used it as a demonstration as to how Killingholme station modified its F.2A hulls to conform to the station requirements. As both machines served with the USN at Killingholme, the identity remains a matter for further research. The US cockade has been applied to the hull but the British cockades are retained on the wings. The machine bears the typical Saunders colour scheme and appears to have been in service for some time as the black fin and hull finish is dull.

and half trays of ammunition were fired on the submarine and splashes observed round the conning tower. As the seaplane passed over it, the submarine turned to port at the same time. After circling the position twice, the patrol was continued. The submarine resembled the *U13-16* class. It was about 130 feet long and newly painted grey; there was apparently no lettering on the conning tower which was of square streamline shape. The bow appeared to be raised but no gun was observed. From the description of this encounter it appears that the F.2A was not armed with bombs but was along for protection of the F.3. N4238 received its chance on 30 August when operating from Tresco. With Lt M.C. Fairhurst (pilot); Capt C.R. Stewart (observer) and crew, the seaplane was on patrol when, at 0903, they sighted the conning tower of submarine about seven miles distant. The submarine submerged and at 0910, 2 x 230-lb bombs were dropped ahead of the oil patch left by the submarine. No results were observed. W/T signal was sent to all ships, and a flare dropped to mark the position.[73]

Another typical attack was that of 3 June 1918, when Houton Bay Station seaplane N4247, under Capt H.A. Wilson, sighted a submarine 12 miles East of Dennis Head. A Very's light was fired from the seaplane, but no reply was received from the submarine that was submerging rapidly. Two 230-lb bombs were dropped, which exploded 25 and 10 yards respectively short of the white swirl made by the submarine – which had completely submerged before the bombs exploded. As a result a quantity of oil and air bubbles rose to the surface, also timber was seen floating on the surface in the centre of the oil patch. The seaplane remained in the vicinity for 40 minutes, but nothing further was seen.[74] N4247 had been delivered in April 1918, and had attacked a U-boat on 20 May without success. It was still at Houton Bay with No.306 Flight in January 1919.

The F.3 operated from Malta with Short floatplanes. They flew from Kalafrana. The Malta seaplane station, along with Gibraltar and Egypt, never had a full complement. Of the 50 officers that were assigned to Malta, throughout 1918 the strength was only about 20. A major cause of this was the denuding of the RNAS strength to go to the operation of the Ottanto Barrage. Gibraltar never receiving its proposed squadron of F.3 boats.[75]

Above & Facing Page: F.2A N4545 has no cockpit canopy. Painted in a dazzle scheme, reportedly red and white, these photographs were taken on 10 October 1918.

Lighter Operations

Porte had made a suggestion as early as September 1916 that the radius of action of the flying boats could be increased by towing them to sea from where they would take off for their patrols. He proposed to use lighters as the method of carrying the large flying boats. The lighter was a channel-shaped vessel that had flotation tanks that allowed it to be partially submerged to allow a boat to enter the chamber. The water would be forced out by compressed air and the lighter would rise bringing the boat out of the water. The lighter was designed to be strong enough to be towed by a destroyer at 25 knots.

On 5 October two representatives of the DNC visited Felixstowe and a design was immediately commenced. Steel construction was adopted so as to use part of the hull as an airtight trimming tank. It was decided that only the stern section of the lighter would be built with a large trimming tank aft so as to submerge the aft end, the flying boat being hauled in by means of a winch. The weight of the flying boats to be accommodated at that time was 4½ tons. The form of the lighter was arranged with a chine for towing at high speed, and made very flat aft, and sufficiently V-shaped forward to enable her to surmount seas without pounding severely. The beam was made sufficient to house a flying boat and to provide side decks on each side for working purposes. No rudder was used and in order to keep the lighter on a straight course three plate skegs were fitted under the bottom aft. The design was tried in the tank at the Admiralty Experiment Works at Haslar and the modified design that resulted from the tank tests was adopted. It had initially been determined to fit greased wooden ways on the slipway and to haul the flying boat directly up these by means of the winch. Owing to the fragile construction of flying boats the decision was changed to use a cradle on a trolley running on rails. The flying boat's step caused problems in that it would foul any support abaft the step, and a special auxiliary trolley was hinged to the main trolley that would hinge down automatically out of the way when the trolley was in its aft-

most position.

Although the order was placed at the end of January 1917 with Messrs Thornycroft of Southampton for four lighters, it was not until 18 June 1917, that the manufacturer carried out a successful test at Calshot with a flying boat being hauled up and relaunched. Towing trials were conducted on the Solent during July and were very successful. The flying boat was successfully towed at 32 knots, showing that the lighter could be towed in any sea that the *America* boats could get off in, and 25 lighters were ordered immediately, the number later being increased to 50 to accommodate H-12, F.2A and F.3 boats, before the further trials were undertaken in the North Sea. There were slight modifications made in order that the lighters be able to take the new F.5 boat. The first was delivered from the new Government shipyard at Richborough where they were constructed by the Royal Engineers. The first was not delivered until May 1918 and only 31 had been received by the time of the Armistice. A further five were completed and the remaining 14 were cancelled.

The lighters had an overall length of 58 feet; beam (outside of plating) 16 feet; depth (keel to side of deck amidships) 7 feet.[76] In February 1918, standing Orders were issued that when a patrol was required that was outside a radius of 150 miles from Felixstowe, the lighters and a supporting naval force were to undertake this task.

Early in 1918 an attack was planned against the U-Boats at Zeebrugge and Ostend. It was important that the German not move ships to fall on the rear of the British attacking

Above: A poor but interesting photograph showing a beached F.2A that has what appear to be pontoons under both wings to support the machine. A motor launch is visible in the background suggesting that the machine has had a poor alighting and had to be recovered before it sank.

Above: Judging by the helmets on the crew handling this *Large America* N43XX, this photograph was probably taken in the Middle East. Note that the boat has two bomb racks under the lower starboard wing.

Above: H-12 8681 on board a lighter.

force and the Felixstowe flying boats were asked to make the reconnaissance flights to ensure that the Germans were not aware of the proposed attack and making any preparations to counter it. For this flight the lighters were used. The three best machines were chosen and "an electric heater, to keep the oil warm, was clipped on beneath each engine, and thick padded covers fitted, to keep the heat in, so that the engines would start easily." When they were shoved down the slipways on 19 March they were picked up by a motor launch that took them to the stern of their lighter. "The five men in the crew of each lighter had flooded the water-tanks in the sterns, and the boats were quickly floated into their cradles and hauled up by a winch into position and secured. With a hiss the compressed air was turned into the tanks, the water blown out and the lighter rose into towing trim."

When they reached their takeoff point the boats were launched, the accompanying ships formed a protective circle around the three boats in case a U-Boat was lurking in the area. Once off Terschelling Flt Lt N.A. Magor in F.2A N4282 turned his formation towards the Bight. At Borkum they ran into two two-seat German floatplanes. Attacking the first, Ens Potter's fire caused it to burst into flames and crash into the water. Flt Lt C.J. Clayton[77] failed to hit the second machine that escaped and alighted at Borkum. The trio did not escape unscathed. The observer in the machine shot down had riddled Magor's boat, the fuel tanks being holed, fortunately above the level of the fuel, and the water pipe on the port engine was hit causing Magor to shut down that engine and proceed on the other. They proceeded on their task while Magor's engineer, Anderson climbed out on to the wing into the 60 knot wind and, over the next hour, repaired the pierced pipe. Now the reconnaissance flight had achieved its objective, they headed for home.

Nineteen days later a second reconnaissance was made using the lighters. It was ascertained that the Germans were not making any serious efforts to clear a passage through the mine-fields, so they were not aware of the proposed attack. Several successful flights were made and it was thought that the machines could be used to bomb German bases in the North Sea.

The USN proposed to use lighters to tow their bombed-up H-16 flying boats from Killingholme to within striking distance of the German submarine pens.[78] In June the Commander of USN Aviation Forces, Foreign Service, Capt Cone, was advised that with regard to the USN's plan to bomb submarine pens "the British are losing interests in this lighter proposal...we will be informed officially that long distance bombing from lighters is not considered practicable and that, therefore, they suggest the abandonment of the project....I think that they will suggest, however, that we use our machines and lighters for long distance reconnaissance similar to the flights which they have been conducting themselves from Felixstowe."[79] The British abandoned their plan to bomb the bases in July.

The British used the lighters, escorted by naval forces, to transport the flying boats if the chosen area for reconnaissance was outside of a 150 mile radius of Felixstowe. Inside that radius they operated from their shore stations. The first

Above: An H-12 aboard a lighter that has a forward cabin. These were used for maintenance of the boats when away from base. Compare how far the rear of the lighter is in this photograph with that of 8681.

reconnaissance took place on 12 March. Flt Lt N.A. Magor[80] was in command. The F.2A boats flew around Terschelling into the Bight where they observed minesweepers at work. German seaplanes from Borkum rose to intercept and Magor's second pilot sent one down in flames, and the other was forced to return to base. Magor's boat had been hit and he had to fly for an hour on one engine while his engineer climbed out on the wing and worked on repairing the water pipe that had been pierced during the fight. After a flight of five and a half hours the formation returned safely to Felixstowe. These long distance flights were not only an exercise in deception to convince the Germans that British intelligence was obtaining its information from these reconnaissances and not from code breaking wireless intercepts, but were important for establishing the location of the ever changing mine fields and activities of German shipping.[81]

On 16 May USN aviators Ens Jay Schieffelin and Ens Benny Lee were taken aboard the destroyer HMS Redgauntlet, together with their crew and two ratings from Felixstowe. Their flying boat was astern in a lighter, and they were joined by two other destroyers towing lighters. Leaving Harwich Harbour in that evening, the three boats were set afloat from their lighters early on the following morning when dawn was breaking. The boats circled about on the water in order to warm up their engines when a Zeppelin appeared. Unfortunately the destroyers opened fire warning the airship of the presence of hostile forces. The three boats took off but "she reached more than 18,000 feet, when we reached our ceiling of 10,000 feet." Giving up any hope of attacking the Zeppelin the boats then carried out their orders photographing any shipping in the Bight of Heligoland. Schieffelin's crew photographed two sailing vessels that they assumed were fishing, but later they were revealed to have been laying a new German minefield.[82]

These flights using lighters to take the boats close to the enemy occupied shore had their share of problems. On 10 August 1918, three F.2A boats (N4300, N4308 and N4531) failed to get off the water and this called the operation of lighters into question. It appears that the Admiralty requested an answer as to why the machines did not perform their planned operation and a Court of Enquiry was held into the incident exactly a month later at Felixstowe.[83]

Major Hallam, the CO of the Seaplane Experimental Station, Felixstowe stated that the weight of the three F.2A boats in August was 11,500 lbs which included six hours fuel at full speed. The original boat was supposed to weigh 7,200 lbs but as supplied from the manufacturer weighed 7,700 lbs. A breakdown in the weights carried accompanied the Court papers and is reproduced to show just what was carried by a war flight. Hallam noted that these

Boats were designed nearly two years ago, since which time Service conditions have demanded additional weights.

Hallam stated that in order to get a boat seaplane off it should weigh no more than 13 lbs per hp. The F.2A was

Above: A *Large America* flying boat behind HMS *Redoubt*.

Below: Aerial view of a *Large America* boat under tow on a lighter. Photograph taken from 10,000 feet on 31 July 1918.

supposed to carry 15 lbs per hp. On this occasion it carried 16.4 lbs per hp. The average load on a Felixstowe patrol was 15.7 lbs with sheltered water to get off in.

Capt E.J. Webster, the pilot of N4308, stated that he was prevented from flying as a man had walked into the propeller damaging it while the boat was still on the lighter. While his boat was otherwise in flying condition he considered it doubtful that he would have got off given the long swell. From his considerable experience "of these Operations machines get badly strained, which affects their ability to get off the water."

Capt A.T. Baker, the pilot of N4351, tried to get off but "owing to the heavy swell, the machine started bouncing so badly that I throttled down." He also considered that taking the boats out in Towing Lighters "does a great deal of harm." Baker also stated that he had taken boats off with 700 to 800 lbs more but this was in the sheltered waters of Felixstowe.

Weight of F.2A Flying Boat for Lighter Patrol 10 August 1918 (in Pounds)

Guns & Ammunition	
4 Lewis Guns	72
14 double trays	126
	198
Stores	
4 Pyrene fire extinguishers	16
Red Cross Box	1
Emergency rations	3½
Drinking water	3
Food in tin box	2½
Pigeons in basket	4
2 sea anchors	8
Tool kit	21
Spare magneto	9
Camp stool	21
	92
Wireless	
Transmitter	21
Receiver	5
Relay	11
Accumulators	33
Tapper	1
Winch and Line	11½
Aldis lamp	5
Aldis lamp battery	21½
Cabinet and wiring	10
Wireless mast & Wires	15
	136
Minor Armament	
Cabin Mounting	2
Revolver & Ammunition	5
Verys Pistol and Cartridges	20½
Bombsight & Release gear	14
Gallows mountings	10
Scarff rings	52
Deflector bags	5
Gun spares	3
Bomb racks	22
Lighter fittings	12
	145½
Miscellaneous	
Camera and plates	14
Binoculars	2
Bigsworth Chart Board	5
Extra water for radiators	42
Code book	2½
Engine chart book & Pad	2
Cockpit cover	2
	89½

The pilot of the third boat (N4300) was Lt J.P. Barnes.

I tried to get off the water, but owing to the heavy swell, I could not get off.

My machine and engines were in good flying condition when I left Felixstowe.

After returning to Felixstowe I found my tail was out of truth: whether this was due to being towed on the Lighter or caused by trying to get off I cannot say.

A wingtip float and end of plane were damaged when getting off the Lighter.

The Station Engineering Officer, Lt F. Susans considered that taking the boats out in Lighters affected the engines due to the constant pounding of the Lighter on the swell would "tend to aggravate any inherent defect, especially applying to water jacket leakages."

The Court found that the swell and absence of wind prevented the boats from getting off. While it may have been possible for the F.2A boats to have taken off with a less load, considering the operational requirements, the distance from base and the threat of enemy aircraft, it was essential to carry the load as detailed. This load was an average increase of 1.050 lbs on a design load from two years prior. "This design apparently did not take into consideration modern War conditions in No.4 Operations Group, which entail long flights combined with offensive action against Hostile Aircraft; also increased weight of W/T apparatus and Hull of Production Machines."

It was recognised that long journeys in Lighters appeared to damage and reduce the efficiency of the F.2A boats. It was recommended that the seaplanes fly from their base and be picked up, if required, after carrying out their duties, at some easily recognised position such as a Lightship within 35 miles of the Dutch coast. Also an improved type of seaplane was required.

The file notes that Rear-Admiral Harwich had proposed to abandon as normal practice the towing of seaplanes by Lighters. The suggestion that the Lighters meet the flying boats for the return journey was felt to be better "although involving a greater risk of losing the aircraft."

In the Air Ministry's letter to the Admiralty reporting on the enquiry it was noted that "Seaplanes of improved design are now under construction, but it will still be necessary, as it always has been necessary, to allow for extra weight to be added over and above the designed weight, as it still seems impossible to prevent the various authorities concerned from overloading machines."[84] Arthur Longmore (later Sir A. Longmore, Air Chief Marshal) wrote that the lighters were not very successful when there was a "beam" wind.[85] J.L. Gordon, recorded that, in his opinion, many of the failures that occurred with lighter operations were due to overloading of the boats, and that "a larger and much more sea-worthy boat than the F.2A or F.5 is essential."[86]

Weight Comparison Summary			
Serial	**N4308**	**N4300**	**N4531**
Machine Empty	7,849	7,774	7,778
Petrol – 250 gallons	1,775	1,775	1,775
Oil - 20 gall	200	200	200
Crew – 5 men	900	900	900
Guns & Ammunition	198	198	198
Stores	92	92	92
Minor Armament	145½	145½	145½
Wireless	136	136	136
Miscellaneous	89½	89½	89½
Consuta Tail	56	56	56
Soakage, etc.	100	100	100
Semaphore	-	33	-
Total	11,541 lbs	11,499 lbs	11,470 lbs

Above: The USN proposed to use lighters to carry their Curtiss H-16 boats across the North Sea until they were in bombing range of the German submarine bases. The anti-submarine war had top priority for material and Liberty engines. This was to lead to the USN and Marine Corps undertaking land based bombing operations and caused conflict with the Army. This photograph was probably taken in the US as the Bureau No. is marked on the hull. The Bureau No. appears to be 3312 but this number was allocated to a US D.H.4. Note the lighter bears the stencil "USN-22" on its bow. The lower part of the radiators appear to be blanked off. (AHT AL0772-156)

Above: A *Large America* boat under tow in a Lighter.

Below: A Curtiss H-12 in a Lighter. Nine personnel can be seen with the flying boat, an indication of what was needed to operate the boats under these conditions.

Above: A *Large America* under tow in a lighter by HMS *Torrid*. When towed at speed in anything other than mild conditions the boats suffered minor faults. This method was one way of extending the range of the *Large America* flying boats and performed a useful service before it was abandoned as related in the text.

Below: Dazzle painted N4296 being loaded onto a lighter at Felixstowe. The striped forward hull could be red or black and white. Note the lack of a cockpit canopy and the small serial at the rear of the hull. The lack of a washboard indicates that this boat has had the rear fabric replaced by ply. A 230-lb bomb is under the wing. N4296 was deleted in the W/E 12 December 1918.

Felixstowe proposed that a force of seven *Large America* boats be made available to attack destroyers off Texel. There is no date on the surviving document so it is impossible to place it accurately in the history of the Large America boats. The surviving Report notes that seven boats would probably be all that was available to Felixstowe in the next two months. To raise a force as contemplated, a percentage of machines would have to be laid up and prepared. The amount of fuel and bombs required would have to be determined. "On the long flights recently made to Terschilling and the North of Holland extra fuel has been carried in tins and the machines would be overloaded if bombs were added."

It was noted that the boats were large machines and would present a good target when attacking a destroyer from a height offering the probability of a good hit, and one or more machines would possibly be brought down. "In view of the small number of these craft available for all purposes and the comparatively small chance of finding enemy destroyers in the area which can be searched with the fuel range available when bombs are carried, it is inadvisable to devote the whole of the Felixstowe force to such an operation for the present."[87]

On 27 February 1918, 36 H-12 of all types were in commission with the RNAS. By 31 October 1918, the RAF had 12 H-12 and 6 H-12 Converted *Large America* boats in service. By contrast, 53 Felixstowe F.2A and 96 F.3 boats were in service. The latter two were still referred to as *Large Americas*.

Actual deliveries of large flying boats were considerably below stated requirements. This was due to technical difficulties and to the fact that special shops had to be erected for production. It was hoped that these problems would be overcome in the next three to four months.

The importance of the *America* boats is highlighted by a proposal of August 1918 for eight squadrons of large flying boats – each squadron comprising 10 machines and three squadrons of floatplanes – each squadron comprising 18 machines, to be in service in the Mediterranean by September 1919. The following was the proposed allocation of these flying boat squadrons:

(a) One squadron at Gibraltar.
(b) One squadron at Bizerts.
(c) Two squadrons at Malta.
(d) One squadron at Taranto with a refuelling base at Cotrone.
(e) One squadron disposed between Otranto and Brindisi with refuelling stations at Corfu, Argostoli and Navarino.
(f) One squadron at Suda Bay with refuelling base at East Crete and Milo.
(g) One squadron at Alexandria with refuelling bases at Sollium and Port Said.

It was anticipated that a large number of machines were to be shipped to Egypt by air and the three stations (e), (f) and (g) would be able to escort these aircraft in lieu of asking for surface craft as well as being well disposed as to hunt submarines. This proposal was considered as good a disposition given present circumstances.[88] The *America* boats were the only types available in sufficient numbers to start such a policy. Whether the Phoenix Cork or other experimental boats would have replaced them later will never be known, fortunately the Armistice ended the need for these flying boats.

By March 1918 it had been concluded that the 230-lb bomb was "scarcely efficacious" in attacks against submarines and the use of a 500-lb bomb was proposed.[89] In April the Air Ministry acknowledged that all twin-engined flying boats and land planes employed on anti-submarine patrols were to be fitted to carry 500-lb bombs. A greater load meant at a larger more powerful flying boat was required. If the war had continued into 1919 the F.5 would have been the boat to have continued the Felixstowe tradition with it being supplied from British, American and Canadian manufacturers. It is also possible that the Fury triplane would have been manufactured in quantity if it had proved capable of operating in the open in the North Sea.

The H-12 in the USN

The USN ordered one H-12 in order to test the type's possibilities. A-152 was possibly the prototype and was fitted with 200-hp Curtiss V2-3 engines mounted behind circular radiators. Its life with the Navy was short as it was destroyed on 30 October 1917, while being prepared for trials. Lt B.L. Leighton blamed the machine claiming it was not airworthy but the Survey Board blamed bad airmanship and inexperience. The USN ordered 24 H-12B type flying boats and spares. The H-12B was "designed for the mounting the 275 H.P. Rolls-Royce engines,"[90] but events conspired to limit the acquisition of only 19 more of the type (Bureau Nos. A-765 to A-783) under Contract 625. They were designated H-12B and powered by 330-hp Liberty engines.[91]

The Curtiss boats were used at Pensacola for training and some patrols were carried out from Hampton Roads such as that of A-770 on 16 June 1918. With crew comprising Lt (jg) H.N. Slater, pilot; Lt H.W. Hoyt, observer; and Ens S.N. Hall, assistant pilot; and a mechanic (who did not get his name recorded), the flying boat left base searching for a dirigible that had been lost from Cape May. The machine flew to Cape May where it landed, took on fuel and oil and repaired a leaking radiator, then returned to base. The patrol was carried out successfully the only trouble being with the failure of the radiator. The Liberty motors gave no trouble, and the distance covered was approximately 385 miles. No record of an attack on a submarine by a US based H-12 has been found to date.

Killingholme, on the east coast of England, was one of the leading naval air stations as it was within striking distance of enemy submarines in the North Sea. It was also central to Allied convoys passing along the coast. Originally an oil supply depot, it was not an attractive town with its houses over-shadowed by the oil storage tanks. It was from here that the USN would make its first combat patrols in British Curtiss H-12 and Felixstowe F.2A boats. The US crews first flew with British crews until they were proficient in the operations carried

Table Showing Aircraft and Seaplanes Required for 1918 Program & Actual Strength on 1 July 1918

Type	Description	Fleet & Carrier[b]		Northern Patrol		Anti-Sub		Special Service		Dunkirk		Mediterranean[a]		Total	
XXX	Large Flying Boat			36	**13**	90	**76**	36		—		39	**5**[d]	201	**94**
XXXI	Float Seaplane	10	**22**	—	**5**	180	**226**	—		—		76	**105**	266	**358**
XX	Ship Fighter	52	**97**[b]	—		—	—	—		—		—		52	**97**
I	Land Fighter	—		—		12	**13**	—		54	**48**	54	**65**	120	**126**
XXI	Ship Reconnaissance	35	**45**[b]	—				—		—		—		35	**45**
III	Reconnaissance Aeroplane	—		—		6	**72**	—		18	**52**	18	**61**[c]	42	**185**
VIII	Light Bomber	—		—		30		—		18		54		102	
XXII	Torpedo planes	36	**0**	—		—		—		—		—		36	**0**
	Heavy Bombers	—		—		—		—		6		—			**9**
Totals		133	**164**	36	**18**	318	**387**	36		96	**100**	241	**236**	854	**914**

Plain figures are machines asked for in 1918 policy. **Bold blue** figures represent strength on 1 July 1918.

Notes: a. Not including machines in transit.

b. Not including aeroplanes of similar type for practice, nor S.B.3D.

c. 36 more in transit.

d. To be fitted from Malta Dockyard production.

Source: RAFM J.M. Bruce Collection Box 59.

out from Killingholme. The base was officially handed over to the US on 20 July 1918.

30 May 1918, saw Curtiss H-12, British serial N4336, make the first all American patrol from Killingholme under the command of Lt C.T. Hull, USN, with Ens A.W. Hawkins as second pilot. They were to meet and escort the USS *Jason* into the Humber River. Lt Comdr Whiting and Lt Leighton together with personnel and tons of material and supplies were on board. The *Jason* was sighted about 40 miles off Filey and accompanied into harbour.[92]

On the night of 9-10 July 1918, N4336, with crew comprising Ens John Jay Schieffelin; Ens John Fanz Staub; E2c(r) Phillip E. Rollhause (W/T); LMM B. Howard M. Ernstein (engineer), was flying ahead of a slow convoy southward bound for London, when a dark streak of oil was spotted on the surface of the water, pointing in the direction of the convoy. Schieffelin thought that it might be from a sunken ship, however Staub indicated that they should attack. The boat dropped Very's lights (flares) to indicate the position to British ships in the vicinity. After a careful run in, two bombs were dropped by the boat. They then descended to 200 feet to watch results.

The six British destroyers escorting the convoy approached in a wide V formation at full speed. "They were coal burners" recalled Schieffelin, "and black smoke gushed from their funnels, above orange flames. I then felt a distinct jar, and directly in front of us rose a wall of white water that looked as solid as a cliff. Putting our flying boat into a vertical bank and executing a "split-S" turn kept us from running into this white wall, and we climbed back to our proper altitude of 1,000 feet." The white wall was the result of the destroyers having simultaneously dropped their depth charges! The aircraft was lucky not to have been wrecked by the explosion!

The oil slick grew larger in size and bubbles appeared. A destroyer stayed at the site overnight and the next day a periscope was seen and fired upon. The submarine was later reported to have made it back to Germany.[93]

N4336 was lost on 19 July while on submarine patrol. It left Killingholme at 3.45 am under the command of Ens Ashton William Hawkins. About 6.45 am on the return course the engineer reported failure of the petrol supply due to an air lock. Two wireless messages were sent before the fuel failed at 7 am. A safe alighting was made in a big seaway. Pigeons were released giving the estimated position of the aircraft. The crew were finally able to get the bilge pump to supply petrol and the engines were started and the boat taxied for some three hours. By noon the sea had moderated about 50% so an attempt was made to take off. The boat attained a level attitude and was being pounded the whole time from one wave to the next and flying speed had just been gained when the aircraft broke in two, three feet forward of the step.

The boat was wrecked instantly, the tail sticking straight up. In the next moments the whole wreckage was covered in floating petrol and caught fire. The fire burned down in about an hour leaving one wing float and the central section of the hull just above water. The empty petrol tanks under this section held up the engines and what was left of the wreckage. The four crew clung to the wreckage.

A Sopwith Schneider had seen the wreck just as the fire had broken out as the pilot was seen waving to the survivors. A Short seaplane was seen flying about four or five miles away but did not see the survivors. A Schneider floatplane reappeared and dropped a life belt before flying off to try and inform the Short as to the location of the wreck without success. The Schneider then circled until it managed to bring a trawler to the rescue of the survivors. The time was about 3 pm. The first trawler picked up the crew and the second got lines on the engines. Both returned to Spurn. The crew were returned to the station by motor launch at about 2.30 am on the 20th. In the week ending 5 September, N4336 was finally officially deleted.

Schieffelin played a part in a dramatic sequence of events on the evening of 18 July. Jay and his co-pilot and navigator, Lt (jg) Roger Cutler, were ordered to report to the station CO, Commander Whiting. It appeared that a submarine had been attacking a convoy using its deck gun. Whiting estimated where the submarine would be expected to be the next morning to recharge its batteries. His orders were to "Search that area at dawn".

With full crew, comprising E3c(r) John E. Taggert (W/T) and LMM B. Howard M. Ernstein (engineer), the pair left in their flying boat the next morning to the area determined by Whiting. While flying over the cliffs on the north side of Flamborough Head at about 400 feet, the air was very bumpy and the flying boat bounced so sharply that

the 360-lb bomb suspended under our starboard wing caused the metal bar under which it was hung to bend enough to release the bomb from the two fork-shaped steel half-loops that held it steady. The bomb was teetering from side to side, and the small propeller (which armed the bomb after release) dropped away, so the bomb was fused and ready to be detonated by a jerk of any kind. I signalled to Roger to release the bomb. It exploded in very shallow water, and the concussion gave us another sharp jolt.

During these adventures the wireless aerial had carried away and they now had no way to communicate with the station.

The sun was just clear of the horizon when we approached Commander Whiting's spot…we sighted northwest of us a submarine, motionless on the surface. Heading for her we made our bomb run. Before we were over the sub, she began to move ahead, and after Roger released our single bomb, I banked the plane sharply so that we could watch its effect. It exploded under the stern of the sub (the bombs were set to explode below, but close to, the surface), and we saw the sudden white foam spot that marked its explosion. The submarine's stern, with its large four-bladed propeller turning briskly, rose clear of the surface, while the submarine slid down under the waves in a very steep dive.[94]

They signalled a trawler by Aldis lamp of the position of the submarine before returning to the station.

The submarine was not put out of business, however, for she later in the day attempted to attack a convoy, but was repulsed by a motor launch which dropped depth charges, forcing the submarine to the surface. In the fight which ensued, (the destroyer HMS *Garry*)... *rammed the submarine* (the *UB 110*) *which was unable to submerge, and after being again rammed, she capsized and sank.*[95]

Schieffelin was told that a German officer and three crew escaped through the gash the *Garry* had made in the submarine's hull. They reported that the submarine could not submerge because their diving planes had been damaged by a bomb dropped by a seaplane.

The Americans also operated stations at Queenstown (commissioned 22 February 1918), Wexford (2 May 1918) and Lough Foyle (1 July 1918) in Ireland from which stations they operated Curtiss H-16 flying boats. Despite the small time frame they operated these stations did valuable work. A brief outline of the history of Lough Foyle serves to demonstrate this.

Work on Lough Foyle Station was commenced about 1 January 1918.

22 July. First seaplanes assembled.
22 August. First practice flights inside Lough Foyle begun.
03 September. First patrol.
06 November. Last patrol.
10 November. Last day of operations.

There were 31 days out of 71 when flying was at all possible. Despite inclement weather many patrols were carried out in rain and gales.

Total number of patrols undertaken: 27.
Number cut short due engine trouble: 5.
Longest patrol: 6 hrs 5 min on 24 October.
Convoys escorted: 10.
Enemy submarines sighted and bombed: 1 on 19 October.
Suspicious oil patched bombed: 2.
Approx 84 hours of patrolling carried out.[96]

Operating the America Flying Boats

Operating these large flying boats called for new thinking in many ways, not the least being their sheer size. As Sir Austin Robinson recalled:

The early seaplane stations were located to suit the needs of aircraft of 1913 or thereabouts. Many of them proved to be very unsuitable when it came to flying boats that needed a much longer run for take-off. South Shields, for example, was a perfectly satisfactory station as long as all that one was flying was Shorts or Sopwith Babys. But the run for take-off was only adequate for flying boats when the wind was fairly strong in the East and very inadequate when the wind was in its normal place in the South West. When I was testing flying boats there,

I always refused to take them off inside the harbour in a South West wind because it involved my flying an aircraft that had never flown before straight up the main street of South Shields.[97] *I had suffered myself from an engine that cut out when I was 100 ft. in the air and I was not prepared to kill a large number of the citizens of South Shields. That station had to be written off as a flying boat station.*

Going further north, Holy Island was fine at a high tide but very inadequate at low tide. Killingholme itself had plenty of water at all times but could be very difficult when the wind was heavily against the tide. Yarmouth one knows all about from the stories of people bouncing boats off there. It could be out of action when the wind and sea were adverse. Felixstowe was perfect. Calshot was all right in normal conditions. They could get a big swell at Cattewater.[98]

A Report on the H-12 boats by Ralph James Hope Vere[99] in March 1917 noted that the

machines are standing up moderately well but have a great deal of trouble with leaky tails. The trouble appears to be that lying in water all the time softens the wood to such an extent that wherever screws are used instead of rivets in fixing planking to floors the screws pull out when getting off the water due to suction on the tail. I am replacing all screws with rivets and hope this will be a great improvement.

However so far we have not lost any machines although we have had two very near squeaks. 8656 was reported by the watchman as sinking at 7PM one night and by working all night and next day we managed to beach her; she is now repaired and ready for work again.

8652 landed at Newly the other day to take Cull to Brest and get her tail planking all pulled out due to the above reasons; however she was beached and has been dismantled there for transport here as soon as I can get a trawler.

8653 (the fourth boat) has not yet arrived as she broke down on the way here and is now at Calshot repairing.

The planes are standing up to the weather wonderfully and so are the engines which, as a rule, start easier than when they lived in warm sheds.

Hope Vere then noted how much rain they had "nearly every day," and very high winds one registering "57 M.P.H. on the anemometer."

On the whole the experiment bids fair to be reasonably successful. If we could only get the Felixstowe type hulls instead of American hulls I think with four machines, I could guarantee 10 patrols a week, given reasonable weather.

With Felixstowe hulls and a well protected anchorage he could not see any reason why the flying boats could not be moored out for an "indefinite period." They had carried out eight patrols of an average duration of three hours. No submarines had been seen, plenty of merchant shipping, but very few patrol boats of any kind.[100]

Difficulties with the reduction gears of the Rolls Royce engines involved an overhaul every 30 to 36 hours and took from two to three days.[101] This caused problems with the numbers of ratings available to carry out this work at Katfirth, Stemness and Strathbeg. In the case of Stemness and Strathberg it was considered possible that special facilities such as slipways in the banks could be provided. The machines could be brought in when due for overhaul and a light scantling being erected to give access to the engines and to protect the engines while undergoing overhaul.[102]

Sir Austin Robinson recalled that the

early boats were all enclosed, so the pilot had no backward view. From about September 1918 most of the new F.2As were open boats somewhat faster in consequence, and with a good view for the pilot. At the same time, they had mahogany tail planking (Consuta sewn planking in the Saunders boats) instead of the narrow washboard and fabric which had given so much trouble at stations where sea conditions were less favourable than at Felixstowe.

They also, from about the same date, had horn balanced ailerons, making them somewhat less tiring to fly than the H-12s and early F.2As; to fly one of those for eight hours with no balance anywhere in the controls, and usually with considerable tail-heaviness half-corrected by a rubber cord was an extraordinarily tiring job and pilots were apt to fold up after a few months of it.[103]

The work of launching a boat was quite complicated, could be very dangerous, and required a large handling party. Pix Hallam described the process of launching a Curtiss boat in *The Spider Web*:

The working party of twenty men gathered around Old "61 and rolled her out of the shed to the concrete area. Here they chocked her up under the bow and tail with trestles in order to prevent her standing on her nose when the engines were tested. Two engineers climbed up to each engine and started them. After they had run slowly for about 15 minutes in order to warm up the oil, they were opened out until they were giving their full revolutions, the tremendous power shaking the whole structure of the boat.[104]

The boat was taken out of the hangar on its beaching trolley and the engine started up and ran until they were running satisfactorily. The four-crew members then joined the boat. The beaching party knocked out the chocks and allowed the trolley to run down the slipway into the water. Here six waders were waiting. They wore watertight breeches that came up to their armpits. Their soles were weighted with lead to ensure they would get a secure foothold. Some waders had been swept away by running tides and were drowned.

Once in the water to a sufficient depth, the boat floated free of the trolley and was taxied clear by using the engines. The boat was then taxied out for take-off, turned into wind and the engines were opened full out. As speed increased the boat hydroplaned onto the step and when the air speed indicator registered 35 knots, the pilot pulled back on the control wheel and the boat leapt into the air. If conditions were not perfect, the boats would sometimes spend a long time trying to get airborne.

Once airborne the pilot's ears were soon deadened to the noise of the engines and would not hear them again unless "something went wrong and the note changed".[105]

In the H-12 the pilot and co-pilot sat side by side in the cockpit. The co-pilot could crawl into the front cockpit to use the Lewis gun mounted on a pillar mount, and later on a Scarff ring, or use the bombsight, if the need arose. The wireless operator sat facing forward behind the pilot on the right hand side of the boat. The W/T set could transmit for a distance of 80 to 100 miles depending on atmospheric conditions. When in the air a weighted copper wire aerial was lowered through a tube that ran through the hull. This could not be used on the water, so a telescopic wooden mast about 30 feet was raised. An aerial was led from the bow, tail, and ends of the upper wing to the tip of the mast. In this way a message could be sent for a distance of about 30 miles. The wireless operator also coded and de-coded all messages. The codebook was bound with lead in the covers so that it could be dropped over the side in the event of capture. An Aldis lamp was available for signalling with ships and other flying boats, emergency rations, the first aid kit and the cage for the carrier pigeons were all the province of the W/T operator.

The engineer was situated in the centre of the boat surrounded by the petrol tanks. His duties were to keep an eye on the engines and take care of the petrol system. Petrol was pumped from the fuselage tanks to the gravity tank situated in the top wing by two wind driven pumps mounted on the hull. The engineer had to regulate the flow of petrol such that the fore and aft trim of the boat was maintained. If an engine had trouble he had to climb out onto the wing and make a repair if possible. He also manned the rear guns in the event of an attack by hostile aircraft.

The Curtiss H-12 did not have a dual cockpit and the 1st pilot had to leave his seat to allow the 2nd pilot to take over. If the weather was foul then the 1st pilot had to decide whether to risk the changeover or continue to fly the boat. The F boats had dual controls so this was not a problem. They also had a better hull and many H-12 boats were converted to F.2A standard by replacing the Curtiss built hull with a F.2A hull to become H-12 Converts.

S.T. Freeman recalled that

Taking-off in an F-boat was a lengthy and strenuous business: the control column had to be pump-handled back & forward until the boat got on to the step. After the war some fighter pilots were sent to learn to fly boats; they were amazed at this muscular technique.[106]

An undated USN report from Florida NAS stated that water often broke through the fabric covering the top of the wing tip pontoons on H-12 and H-16 flying boats, allowing it to fill with water. "This makes the boat difficult to set on its handling truck and necessitates a considerable wait until the water runs out the drain plug." Florida suggested a veneer or light ply covering for the pontoons.[107] It is not known if this was adopted. No British report of this fault has been found to date.

Conclusion

Writing in a paper for the RAF Staff College, J.K. Waugh admitted that

by intelligent use of their submarines, the Germans came, in April, 1917, within measurable distance of winning the war. In that month alone the shipping losses through enemy action, principally submarine, attained the appalling total of 881,000 tons. These sinkings represented, besides the great loss in carrying power, the loss of hundreds of thousands of tons of food and munitions of war of which we stood in grave need.

This damage was effected by the use of a limited number of submarines, of comparatively small radius of action, operating from a few somewhat insecure bases, the more important of which were continuously harassed by air and sea attack. Had the enemy possessed safe bases and a few more submarines of greater radius it is quite conceivable that we should have been forced to sue for peace, regardless of the actual military situation in France.[108]

While the majority of those engaged in the endless work of the RNAS and RAF patrols considered that their main role was to sink submarines, in addition the *America* boats performed another valuable task in escorting the "Beef Convoys." These were conveyed at irregular intervals during daylight from the Hook of Holland to Harwich. The work was carried out by relays of two or three machines. They would fly ahead of the convoy, flying a semicircular path, in a radius slightly greater than it was possible for a submarine to intercept the ships. The best height was found to be between 500 to 600 feet. In addition the boats kept a lookout for mines and these were often sunk by machinegun fire from the flying boats.

On 21 July 1917, seven patrols were ordered as a submarine had been attacked on the convoy route the previous day 5 miles south of the North Hinder lightship. Due to boats being in for overhaul, two boats had to be sent out twice. Each did a patrol of 540 miles. They were hauled out of the water, cleaned, refuelled, the machine guns stripped, cleaned and oiled, and the ammunition checked. All control wires were examined and the engines checked. One of these boats on its return from patrol suffered a splintering crash. The Lewis gun drum had come adrift and hit the port airscrew shattering a blade that shot splinters into the hull. Fortunately no one was injured and the boat managed to return on one engine.

Not all these escorts were tedious. Ens Schieffelin was engaged in one such activity from Felixstowe on 5 June 1918, when a "near miss of a tremendous anti-aircraft "Archie" shot that exploded directly ahead of us, gave our flying boat a tremendous bounce." In the misty conditions the destroyer had mistaken their boat for an enemy aircraft. Schieffelin noted that they were really responsible as they had approached from out of the sun as if on an attack course.[109]

"Convoying was tedious, but in retrospect it is gratifying to know that a convoy under escort by aircraft, in this part of the North Sea, was never even attacked by a submarine, while a number of merchant vessels and destroyers were sunk by enemy submarines in this zone, when escort by aircraft was not possible."[110] Sqn Ldr J.K. Waugh, when considering the short sea convoys where air escort was practicable throughout the voyage, noted that from April 1917 to the end of the war, 312 ships were torpedoed whilst in convoy. "Of these attacks only two occurred when the vessels were under air escort. During the period over 8,000 escort flights were carried out."[111]

Another task undertaken by the boats was the location of mine fields. This was extremely difficult in the North Sea and it "was impossible to locate a mined area unless the mines were badly laid or broke the surface, and even then it was difficult, often impossible, to be sure whether they were moored or drifting.... Although a negative report by aircraft is no indication that the area is not mined, these patrols did, however, meet with a certain measure of success, being the chief duty of the flying boats after the armistice."[112]

A post-war assessment of the anti-submarine war noted that the Admiralty had originally specified for a large and a small seaplane having qualities of speed, bomb carrying and endurance for in-shore and long range patrols. "Subsequently in the light of experience with German Seaplanes, a third type – a fast handy fighting seaplane – was also requisitioned." Production of *Large America* flying boats was held to have been retarded by the lack of suitable engines, and when these were available in quantity, housing difficulties and difficulties in the provision of a sufficient number of trained pilots meant that full use was never made of the type. As a consequence, a large proportion of the work was still carried out by the Short and Sopwith Baby seaplanes up to the signing of the Armistice.

This document stated that the reports "rendered by various Group Commanders" noted the following problems with the *Large America* boats. They could not be flown for long in bad weather due to heaviness of controls; they were difficult and expensive to house; and they could not get off the water with a full load after a few weeks service, a condition that was further aggravated when they were left in the open, "in these respects they are inferior to the Short Seaplanes."[113] This report was stressing the advantages of the land based aeroplane over the flying boat, and while bearing this in mind, the statements do bear consideration. Flying boats were more expensive than floatplanes; however, the ability of these machines to operate with ever increasing loads and to survive a forced alighting at sea is well illustrated in the main text. Floatplanes were notorious for their inability to stand up to the open sea. As J.L. Gordon recorded, long patrols over the sea are always dull and tedious. Forced alightings "are usually accompanied by many discomforts which result in loss of enthusiasm. The success of this (seaplane) work in large measure, depends upon the sea-worthiness of the machine and its equipment, and also upon the pilot knowing that in the event of a forced landing, no effort will be spared immediately to locate the disabled flying-boat and rescue its crew." [114] In these respects the *Large America* boats proved themselves time and again.

In 1921 the F.2A and F.5, along with the Avro 504K, Bristol Fighter, De Havilland D.H.9 and D.H.10, Sopwith Snipe and Vickers Vimy, survived being declared obsolete with the remaining wartime aircraft types. There were 45 with squadrons, 24 with schools and 40 in store.[115]

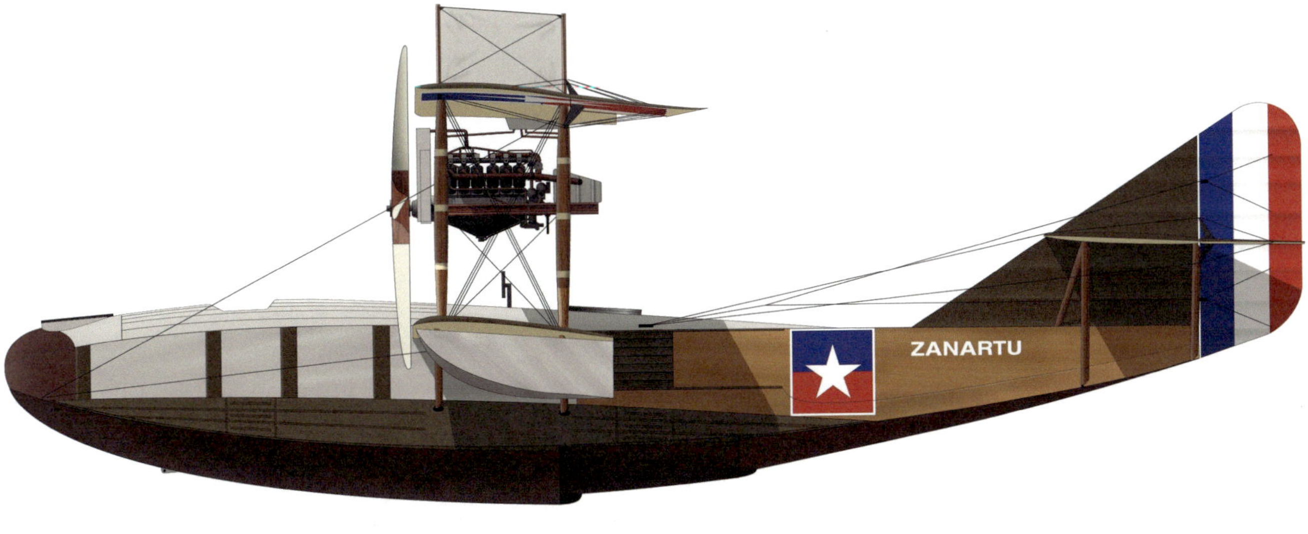

F.2A *ZANARTU* in post-war service with Chile

Chapter 4 Endnotes

1. The Harwich Force was called the Light Force as it was composed of a large numbers of destroyers and light cruisers based at Harwich. German submarines were surrendered at Harwich after the war's end under the supervision of the Harwich Force.
2. Memorandum: Secret: *Seaplanes operating in conjunction with Light Cruisers*, R.Y. Tyrwhitt, Commodore (T), 07.06.1914. AIR1/148/15/87.
3. Log book entries supplied by A. Hellwig. Original Log books are held in the Australian War Memorial.
4. HM Seaplane Base Malta, Secret: Intelligence Report. Week Ending 11th August, 1917. The serial for the F.B.A is given as 4275. This is probably its original French serial number. TNA AIR1/668/17/122/760.
5. Hallam, T.D. *The Spider Web*, P.34. Hallam (later Major, DFC and Bar), was OC of the Felixstowe War Flight.
6. Wing Cmdr J.L. Gordon, DFC, Royal Canadian Air Force. "War Experiences (1914–1918)." Source not stated but thought to be from the series of Lectures and Essays from the Work of the Officers attending courses at the Royal Air Force Staff College, and published by the Air Ministry. Photocopy in RAFM J.M. Bruce Collection Box 61.
7 Hallam, T.D. *Op. Cit.* P.49.
8. Hallam, T.D. *Op. Cit.* P.113.
9. Flt Sub Lt L.P. Paine to CO, Calshot, 25.04.1917. TNA AIR1/661/17/122/640.
10. Chamier, J.A. *The Birth of the Royal Air Force*, Sir Isaac Pitman & Sons Ltd, UK, 1943. P.104.
11. Report from Flt Lt C.L. Scott to CO, Calshot, 26.04.1917. TNA AIR 1/660/17/122/629.
12. Report from Wing Cmdr A.W. Bigsworth, to Director of Air services, 26.04.1917. TNA AIR1/660/17/122/629. The "Report of work carried out by RNAS Units in Portsmouth Group during 1917" records the following entries for 8655: 24.04.17. Flt Lt C.W. Scott; Flt Sub Lt Paine. Sighted sub at 1050, position 15 mls S (true) of Portland, dropped two 100-lb bombs. This sub was apparently damaged & later sunk by a destroyer. TNA AIR1/659/17/122/609.
13. The author has not been able to identify *Perham.*
14. Bob Leckie was a Scots-born Toronto Canadian who joined the RNAS. He graduated from the RNAS flying school at Chingford before being sent to Felixstowe where he was trained to fly the America boats. Post-war Leckie would remain in the RAF and was sent on loan from the RAF to Canada as Director of Flying Operations for the Air Board. In 1940 he was sent back to Canada to help set up the Commonwealth Air Training Plan. He transferred to the RCAF in 1942 and went on to serve as CAS, RCAF, retiring in 1947. During WWII he was again involved in anti-U-Boat activities. He died in Ottwa in early 1975.
15. Sqn Commander D. Oliver, Gt. Yarmouth to Director of Services, 14.05.1917. TNA AIR1/660/17/122/623. This file also contains the reports of Galpin, Leckie and Flt Lt E. Cadbury who was on the scene in a Short floatplane 8066. S.F. Hollister wrote to J.M. Bruce that AM "Stiff" Goody "became so excited at the time he caught his hand in the blade of one of the 'Wind Pumps.' I remember him giving me a vivid account of the action when he came ashore, how they had climbed and came in the sun's rays onto the top of the Zeppelin." S.F. Hollister to J.M.Bruce, RAFM J.M. Bruce Collection Box 42. Hollister had flown everything from the early *America* boats to the Felixstowe Fury.
16. Confidential letter 25.05.1917. Copy in TNA AIR1/637/17/122/153.
17. Cdr Longmore report on a visit to Felixstowe. Minutes of the 5th Air Department Weekly Progress Meeting, 12.06.1917. TNA AIR1/270/15/226/113.
18. Robert Fredrick Lea Dickey. Dickey "was a little button of a chap, but what he lacked in size he made up in blood-thirstiness." Hallam considered him one of the best second pilots it was possible to desire. "He always gave his first pilot the comfortable feeling of being absolutely trusted, and this was why I liked flying with him." (Hallam, T.D., Op. Cit. P.114.) Dickey was born in July 1895. His father was a professor at McCrea Magee College, Derry. He was educated at Foyle College and was cox of the Derry Eight that won the 1912 Cork Regatta. Enlisting in the RNAS in April 1915, he was selected for pilot training taking his certificate at Chingford in December 1916. He was forced down in June 1918 and interned in the Netherlands. Called up in WWII he attained the rank of Lt Commander in the RNVR.
19. For an overview of the British use of wireless and wireless intercepts see Hezlet, A. *The Electron and Sea Power*, (Peter Davies, UK, 1975). The Germans were conscious of their code books falling into the Allies hands and the Zeppelins were ordered to only take the HVB code when carrying out a bombing mission. The radio signal "only HVB on board" was soon understood by the British that a Zeppelin was on a combat patrol. The British learned to be more careful of their own radio traffic because of what they could learn from radio intercepts. The British established Direction Finding radio stations from 1915 and were able to plot the flight path of the Zeppelins. In fact sometimes they were more aware of the position of the Zeppelins than the Zeppelins themselves.
20. Report of B.D. Hobbs and R.F.L. Dickey, 14.06.1917. The CO of Felixstowe wrote that these "two officers on several occasions have been out on these long Anti-Zeppelin Patrols. They are very keen and courageous to a fault." TNA AIR1/637/17/122/156. Hobbs who already had received the DSC for work against submarines, was granted a DSO, Dickey the DSC and Davis and Goody the DSM.
21. Report by Flt Lt C. Galpin, 26.05.1917. TNA AIR1/637/17/122/148.
22. A7459 and A7457 were equipped with the unreliable R.A.F.3a engine. They were modified with additional fuel tanks and special cameras for a proposed long-range

photographic reconnaissance flight to the Kiel Canal. The increased fuel supply was to provide for a 14 hour endurance. The machines were given a special camouflage of matt fawn and sky blue. The original crews selected were Sqn Cdr, the Master of Sempill; Flt Lt F.S. Cotton; Lt G.S. Trewin, DSC, and Lt E.B.C. Betts. The crews and aircraft went to Hendon to train and were at Bacton for their proposed flight in early August 1917 when the Admiralty apparently called the whole flight off. The two D.H.4 biplanes stayed at Great Yarmouth on anti-airship activities. With their increased tankage they would be able to accompany the flying boats on their long patrols. Photographs show an additional two Lewis Guns were mounted on the top plane centre section and this may have been its configuration on 5 September.

23. Nicholl was to die of an internal complication at 38. His condition had grounded him five years before. For a pen picture of this true English gentleman see Macmillan, N, *Tales of Two Air Wars*, G. Bell & Sons, UK, 1963, P.140. Macmillan thought that the pair of Nicholl and Leckie were "admirably complementary".
24. Recollections of *Oblt* R. Frey, the observation officer of *L46*. Snowden Gamble, C.F. *The Story of A North Sea Air Station*. P.263.
25. Handwritten notes in J.M. Bruce Collection, RAF Museum. The author is not identified. The crew of the D.H.4 were Flt Lt A.H.H. Gillian and Lt G.S. Trewin, whilst that of the flying boat were Flt Lt Robert Leckie (pilot), AM Thompson (W/T operator) and AM Walker (engineer), and with Sqn Commander V. Nicholl in overall command.
26. OC Great Yarmouth to CO Lowestoft, 10.09.17. TNA AIR1/668/17/1222/763.
27. CO, Lowestoft and Great Yarmouth, to Director of Air services, 10.09.1917. *Ibid.*
28. Minute: signed by Hope. 18.09.17. Approved 19.09.1917. *Ibid.*
29. Secret Memo: Charles Walker to CO, Lowestoft, 20.09.1917. *Ibid.*
30. Lt Hope to OC, Cherbourg, 15.05.1918. TNA AIR1/718/28/4.
31. Senior Patrol Officer, RAF Cherbourg, (probably Capt C. Scott), to OC, Cherbourg, 15.05.1918. *Ibid.*
32. Writing to J.M. Bruce in 1960, *Fitz* Cowley recorded that he has met with Pattinson who was then living in the Lakes District, and he "showed me a model which has been made professionally in recent months of F.2A N.4291 which he was flying when he met *L62*. This machine was called 'Black Bess' and was painted jet black... and Pattinson assures me ...the model is 100% accurate." Cowley goes on to record that during "the last war Pattinson was asked by the Admiralty to experiment in a flying boat glider. It was proposed to tow these aircraft to Norway where a lack of landing areas was a problem. So he modified a Slingsby glider fuselage into a hull and fitted Slingsby wings. This was towed off the Lake by a motorboat and flew quite well. The idea was not followed up, however, and the wings were returned to the Admiralty but the hull is in Pattinson's workshop where I saw it." H.F. Cowley to J.M. Bruce, 07.09.1960. RAFM J.M. Bruce Collection Box 24.
33. Albert Henry Munday was born in Melbourne, Australia, but is claimed as a Canadian in Wise's history (*Canadian Airmen and the First World War*). Munday was educated at Eaton and in Canada, and was a reporter for the Toronto *Telegram* newspaper before he joined the RNAS in 1916. Recovering from a crash while with 12 Naval Squadron, he returned to Canada where he wrote *The Eyes of the Army and Navy*, a practical text on all aspects of aeronautics. On his return to the UK he requested seaplanes but got off to a bad start by wrecking Short 320 N1362 on 21 February 1918. He and his observed had to be rescued by HMS Ouse. He studied Journalism at Columbia University post-war. He was to write several more books and many articles and helped establish the Toronto Flying Club. He had two brothers who also served in the RAF.
34. Report by Capt Mundy with letter from F.W. Bowhill, Killingholme to Emonds, 12.05.1918. Copy in TNA AIR1/15/15/1/64.
35. Letter T.C. Pattinson to J.M. Bruce, 03.09.1956. RAFM J.M. Bruce Collection Box 22. Also Sturtivant & Page, (*Royal Navy Aircraft Serials & Units 1911–1919*) P168. *Op. Cit.*
36. Wing Cmdr J. Lindsay Gordon, DFC, Royal Canadian Air Force. "War Experiences (1914–1918)." Op. Cit. For a pen portrait of Gordon see Carr, W.C. *Out of the Mists*, P.28.
37. Norman Ansley Magor and Chas Edward Stafford Lusk were Canadians. Magor was awarded the DSC for this action. The 27 year old Capt Magor was lost with the crew of H-12 8677 on 25 April 1918. Lusk was from Toronto, and learned to fly at the Toronto Curtiss School. He was to rise to the rank of Captain.
38. Abbatiello, J.J. *Anti-Submarine Warfare in World War I – British naval aviation and the defeat of the U-Boats*, Routledge, UK, 2006, P.1. Dodds, R. (*The Brave Young Wings*, Canada's Wings Inc, Canada, 1980.) identifies the fighter escort as Sopwith Camels.
Bombs used varied from 20-lb at the beginning to 500-lb bombs towards the end. The most popular general purpose bomb was the 230-lb bomb. It had a high percentage of explosive to total weight and could be fitted with instantaneous, 2½ and 9 second delay fuzes. The most useful was the 2 ½ second delay that exploded the bomb at about a depth of 40 feet and was effective against submarines on the surface and to a depth of up to 60 feet.
39. Wise, S.F. *Canadian Airmen and the First World War*, P.151. Wise is referring to the air assets in 1916, but this attitude continued until the amalgamation of the RFC and RNAS to form the RAF. For the problem of unescorted flying boats and the impact of the Brandenburg fighters on the British reconnaissance flights see TNA

AIR1/643/17/122/269. In reply to suggestions to increase the armament of the flying boats the Admiralty was concerned that "no permanent alterations to *Large Americas* should be undertaken which would impair their special capability to make long reconnaissance flights."

40. Admiralty to Air Ministry, 05.07.18. TNA AIR1/ 643/17/122/269.
41. *Sinking of Submarines by Aircraft*, Copy in NARA RG72. File A-I-Z.
42. Sturtivant & Page, *Op. Cit.* P.141.
43. This section has been written in the main from Hallam, T.D., *Op. Cit.* PP.94–102.
44. Letter: George Hodgson to Sir Austin Robinson, 18.03.1981. Copy in RAFM JM Bruce Collection Box 22. 3067. Wright had been the observer in 3072 when she suffered a forced alighting on takeoff on 23.01.1917.
45. This incident is recorded in both *The Spider Web* and *The Story of a North Sea Air Station* which indicates the effect is had on RNAS personnel patrolling over the North Sea.
46. Hallam, T.D., *Op. Cit.* P.103.
47. Robinson, A. Letter to Hodgson 04 April 1981. RAFM JM Bruce Collection Box 42.
48. C.F. Snowden Gamble gives the serial incorrectly as N8660. See Sturtivant & Page Sturtivant, R. and Page G.
49. The names used in this section have been accepted as follows as they are different in the following references: Sturtivant & Page give C.T. Young, under the entry for 8660, Hobson's recording has been accepted in this case. Both Sturtivant & Page and Hobson give J.T. Roe. There is no J.T. Roe in Arthur's work (*Contact*, Naval Aviation Register, USA, 1967.) and *Flying Officers of the United States Navy 1917-1919*, and their recording of G.T. Roe has been accepted as correct, as it is the same as that given by the Naval Attaché at The Hague.
    Sturtivant & Page state that only Grant was captured. Hobon records that Roe, Grant and Money were taken POW, Arthur confirms that Roe was taken POW.
50. Minutes of the Air Department Progress Committee of 15.05.17. TNA AIR1/270/15/226/113. The boat would have been H-12 8664 from Tresco that exploded and fell in flames 1 mile SW of Gush Island, Scilly, on 09.05.1917. The crew of Flt Lt John Cedric Railton, Flt Sub Lt Robert Scott Whigham and Ldg Mech David J. Birse were all killed.
51. Felixstowe to Director of Air Services, 04.07.1917. TNA AIR1/667/17/122/745. 8679 had been delivered to Felixstowe on 17.03.17, for erection. Nothing further is known of its career before its firey demise. 8682 had encountered a submarine on 14.06.1917, and dropped four bombs. It was surveyed on 20.08.1917, when the hull was reported as worn out, more probably due to shrapnel damage, and deleted on the 31st.
52. Cole, C. & Chessman, F. *The Air Defence of Britain 1914–1918*, Putnam, UK, 1984. P.255.
53. Robinson, A. *Op. Cit.*
54. US Naval Aviation Forces, Foreign Service, *Confidential Bulletin No.15*, 17.08.18. Also US Naval Forces Operating in European Waters, US Naval Aviation Forces, Foreign Service. USS Melville, Flagship. 01.10.1918.
55. John Jay Schieffelin, known as *Jay*, was born in Scarborough-on-Hudson, New York, on 31 October 1897. He served with the New York Cavalry on the Mexican Border in 1916, entering the Navy Reserve Flying Corps in April 1917 as he was a prime mover of the Second Yale Unit. Naval Aviator on 7 November 1917, was sent to Moutchic, France for further training in January 1918. From February to Armistice he flew long rang patrols from Felixstowe at first, then Killingholme. On the US entry into WWII he reenlisted as a Lt Cmdr and retired in February 1946 as Captain. He had served in the Pacific and Japan.
56. Group Capt Gerald Edward Livock, DFC, AFC. His autobiography *To the Ends of the Air*, was published in 1973. Livock flew Short and Sopwith floatplanes before going onto the *America* flying boats, he also had a long career post-WWI commanding the Flying Boat Development Flight at Felixstowe. He was with No. 205 Squadron, the first RAF squadron to be stationed in the Far East from 1929 to 1931.
57. Snowdon Gamble, C.F. *Op. Cit.* PP.362–364.
58. Harold Mott Wilcox was born on 2 January 1889 in Bethlehem, PA. He obtained an LLN at the Pennsylvania Law School in 1913. He had served in the Pennsylvania National Guard before enlisting in the Navy in July 1917. He was trained at Pensacola before being sent to Killingholme. He left the Navy in July 1921 as a Lt (jg). He died on 2 May 1935.
59. Knapp, M.W. "Flying Boats over the North Sea," *Over the Front Journal*, Vol.9 No.2, 1994. P.128. Snowie (*Collishaw & Company – Canadians in the Royal Naval Air Service 1914–1918*) states that Ellis had broken his legs and that after he left the RAF in March 1919 he received a pension of £200 per annum for life. Knapp's article contains exerts from Wilcox's diary that are very interesting in their description of the British and the operations of the Station. He was not enamoured with the British and wondered why America was in the war. According to Sir A. Robinson the Rolls Royce engines had a tendency to shed propellers into H.12 hulls when the reduction gears seized. Correspondence with Sir A Robinson, Comments on Professor Robinson's Notes: Feb–June 1959. RAFM JM Bruce Collection Box 22
60. Snowdon Gamble considered *Oberleutnant zur See* Friedrich Christiansen a professional seaman, "also a skilled aviator, an excellent shot and a sportsman." Christiansen was to be credited with a Porte Baby flying boat, seven *Large America* flying boats, the submarine *C.25* and the Airship *C-27*. He was the only naval aviator to win the *Pour le Mérite* in WWI. He was to sullen his reputation in the Second World War when he was Military Governor of the Netherlands. Charged by the Allies with war crimes committed by the SS

in his zone he was imprisoned but later pardoned.
61. Copy of Report in *Report of Engagement with Hostile Seaplanes & loss of large America Seaplane No. 8677*. 25 April 1918. From Rear Admiral Hawich to Admiralty. TNA AIR1/342/15/226/273.
62. A German officer's description of the action. Quoted by Snowden Gamble, C.F., *Op. Cit.* P.383. A similar report of three survivors is given for the loss of Curtiss H-12B N4338 on 15 February 1918, with crew of Flt Lt Claude Chester Purdy (another Canadian in the RNAS), Ens Alan D Sturtevant, USN, (pilots); AMI Sidney James Hollidge and 17 year old Boy Mech Arthur Hector Stephenson. At this time the boat had been converted with a F.2A hull. Sturtevant was a member of the First Yale Unit and was the first USN airman to be lost in aerial combat in WWI. See Arthur, R.W. Op. Cit. PP. 531–532 for details of letters obtained from Germany after the war by Sturtevant's father. See also Wortman, M. *The Millionaires' Unit*, P.161.
63. Magor's younger brother, Capt Gerald Atkinson Magor of No. 201 Squadron, was shot down in his Sopwith Camel B6428 on the Western Front on 22 April. Credited to Ltn Hans Weiss of Jasta 11, he was taken POW but died three days later according to Snowie. Hobson (*Airmen Died in the Great War 1914–1918*) records his death as on the 22nd.
64. Quoted in Arthur, R.W. *Op. Cit.* P.68.
65. The first patrol of the day may not be considered typical but it shows the dangers of flying over the sea that were present even when the enemy was not. Patrol Reports for 2 October 1918. Copy of Report in TNA AIR1/165/15/312/42.
66. Recorded as Stevens by Snowden-Gamble.
67. C.F. Snowden Gamble recorded the serials of the boats from Great Yarmouth but they are not recorded in Sturtevant & Page.
68. Part of a letter written by Ens K.S. Keyes, USNRF, contained in a four page report on US Naval aviators operations with the British. No title, no date. NARA GU General Information on Aviation. Box 158.
69. Report from the US Naval Attaché at The Hague. The Dutch authorities asked if the American officers would give their word of honour individually not to escape. They were then allowed freedom to move about and were not restricted to an internment camp. Contained in a four page report on US Naval aviators operations with the British. No title, no date. NARA GU General Information on Aviation. Box 158.
70. Report reproduced in Jones, H.A. *The War in the Air*, Vol. VI, P.362.
71. This victory is credited to *Obflugm* Metzing and *Flgm* Walker. Snowie, *Op. Cit.* P.147. Ens J.J. Scheffelin, USN, took part in the rescue. See Shirley, N. (Ed.) "Wartime Memoirs of John Jay Schefflein," *Over The Front*, Vol.9 No.2, P.99.
72. Samson, C.R., *Fights and Flights*, originally published 1930, Battery Press, USA, republished 1990. P.360.
73. US Naval Aviation Forces, Foreign Service, *Confidential Bulletin No. 24*, 02.11.1918.
74. Patrols. NARA RG45 GU General Information Aviation, Folder 1, Box 158.
75. Wise, S.F. *Op. Cit.* P.216. The resources used for the Otranto Barrage were in the whole wasted and it was the convoy system that was successful in the Mediterranean. There were actually more submarines penetrating the Barrage in the period after it was strengthened than before.
76. This section has been written in the main from "Seaplane Towing Lighters" photocopy from TNA AIR1/2103.
77. Cecil John Clayton was another Canadian who joined the RNAS. He was promoted to Major and took over command of No.230 Squadron on 20 August 1918. Post-war he returned to Canada and studied to be a dentist.
78. According to Wise the attacks were to be a joint Anglo-American bombing offensive against German naval bases. Wise, C.F. *Op Cit.*, P.199 Note.
79. Lt W.A. Edwards, Aide for Aviation, to Capt Cone, Commander, USN Aviation Forces, Foreign Service. 06.06.1918. NARA RG45 Records Collection of the Office of Naval Records & Library, Subject File 1911–1927, GP – Plans & Policies – Aviation Plan & Policies; Press Releases, Entry 520 Box 147.
80. Possibly F.2A N4282 that had an engagement and claimed a two-seater on this date. Sturtivant, R. & Page, G., *Op. Cit.*, P.237.
81. Adm Beatty had been shaken when the German sortie in April had not been picked up by Naval Intelligence, and never thereafter completely trusted the codebreakers of Room 40. Wise, S.F., *Op. Cit.*, P.199, Note.
82. Shirley, N. (Ed). *Op. Cit.*
83. This section has been compiled from the testimony and reports in TNA AIR1/652/17/122/465.
84. W.A. Robinson, Air Ministry to Admiralty, 21.10.1918. TNA AIR1/652/17/122/465.
85. Longmore, A. *From Sea to Sky 1910–1945*, Geoffrey Bles, UK, 1946. P.68.
86. "War Experiences (1914–1918)" by Wing Cmdr J.L. Gordon, DFC, Royal Canadian Air Force. *Op. Cit.*
87. Copy Report in AIR1/305/15/226/153.
88. Minute from Senior Air Officer to CinC, 31.08.1918. TNA AIR1/671/17/134/10.
89. Letter to Admiralty, Secret, 29.03.1918. TNA AIR1/643/17/122/269.
90. W.A. Prentice, Dept of Aeronautical Supplies, to Bureau Construction & Repair, 10.09.1917.
91. The designation H-12C and H-12L have been given for the USN's version of the H-12 in some publications, however it is only identified as the H-12B in the chart *Characteristics of Flying Boats and Seaplanes* that was issued in 1918.
92. Atwood's Log Book gives H-12B (N)4336 piloted by Lt Cleave as "My first boat hop. Whitly looking for the Jason." This may have been an earlier flight, no time given in Log Book. USN Aviation Museum, Pensacola.
93. Shirley, N (Ed). *Op. Cit.* P.114. The Admiralty report of

21.08.1918, states that the destroyers dropped their depth charges before the aircraft released its bombs, however Schieffelin's recollections seem to be more accurate as the boat's crew would have wanted to attack first while the submarine was in sight which was the accepted practice. The resultant explosions would denote the area to surface vessels. Copy in NARA RG45 Records Collection of the Office of Naval Records & Library, Subject File, 1911–1927. Entry 520, Box 143.
94. *Ibid.* P.99.
95. *Flying Officers of the United States Navy 1917–1919*, P.21.
96. Ens GG Zabriskie, AIO, "Summary of Operations of U.S.N. Station, Lough Foyle, 1918." Copy in NAHC.
97. Robinson was posted to South Shields on 5 July 1918, to test and deliver Felixstowe F.3 boats that were being built there. Until he was demobilised in April 1919 he was "flying and testing new flying boats all around the coast and thus handled an enormous number of different boats." Sir A. Robinson to A.B. Massey, 04.11.1980. RAFM J.M. Bruce Collection Box 22.
98. Sir A Robinson to J.M. Bruce, 31.05.1980. RAFM JM Bruce Collection, Box 22.
99. Although not stated in the Report, the location is thought to be the Isle of Tresco, part of the Isles of Scilly, as Lt Cmdr Hope-Vere was in command of the RNAS here. Flt Sub Lt Hope-Vere gained his Royal Aero Club Aviator's Certificate No.900 on 9 September 1914. He participated in Porte's experiments with various hulls at Felixstowe including flying 3569 on 30 May 1916. As Lt Col Hope-Vere, AFC, he flew the F.5 that visited the ELTA exhibition at Amsterdam in 1919.
100. Copy of Report in TNA AIR1/305/15/226/153 Pt.II. See also Walpole, N., "Tresco's Sentinels", Flypast, UK, December 2008, P.30.
101. There were occasions where the reduction gears failed causing the airscrews to come adrift. See Meadows, J. "Diary of A North Sea Patroller," Aeroplane Monthly. September 1997, P.55.
102. YAW to Lt Cmdr Andrews, 15.12.1917. TNA AIR1/305/15/226/153 Pt.II.
103. "F-Boats – a postscript by an ex-pilot." *Flight*, 20.01.1950. P.87.
104. Hallam, T.D. *Op. Cit.*, P.30.
105. *Ibid.* P.57.
106. Interview with W/Cmdr S.T. Freeman. Notes in RAFM JM Bruce Collection Box 24.
107. CO, Pensacola, NAS to Bureau Construction & Repair. NARA RG72, Box 385 File O-ZF-2.
108. "Anti-Submarine Patrolling" by Sqn Ldr J.K. Waugh, DSC. Source not stated but thought to be from the series of Lectures and Essays from the Work of the Officers attending courses at the Royal Air Force Staff College, and published by the Air Ministry. Photocopy in RAFM J.M. Bruce Collection Box 61.
109. Shirley, N (Ed). *Op. Cit.* P.111.
110. "War Experiences (1914–1918)" by Wing Cmdr J.L. Gordon, DFC, *Op. Cit.*
111. "Anti-Submarine Patrolling" by Sqn Ldr J.K. Waugh. *Op. Cit.*
112. "War Experiences (1914–1918)" by Wing Cmdr J.L. Gordon, DFC, Royal Canadian Air Force. *Op. Cit.*
113. *Aircraft v. Submarine. Submarine Campaign 1918*, Vol.1, Part 4 of the Technical History and Index, a Serial history of Technical Problems dealt with by Admiralty Departments, Technical History Section, March 1919. P.5. This Report is prejudiced against marine aircraft, however it was not until well into World War II that land based aeroplanes were developed to the point that they could carry out long range anti-submarine patrols.
114. "War Experiences (1914–1918)" by Wing Cmdr J.L. Gordon, DFC, *Op. Cit.*
115. Tabulation in unknown book, source not stated. Photocopy in RAFM J.M. Bruce Collection Box 59.

F.2A N4546 was a Felixstowe boat. It arrived on 28 July 1918 and was destroyed in a crash on 21 August.

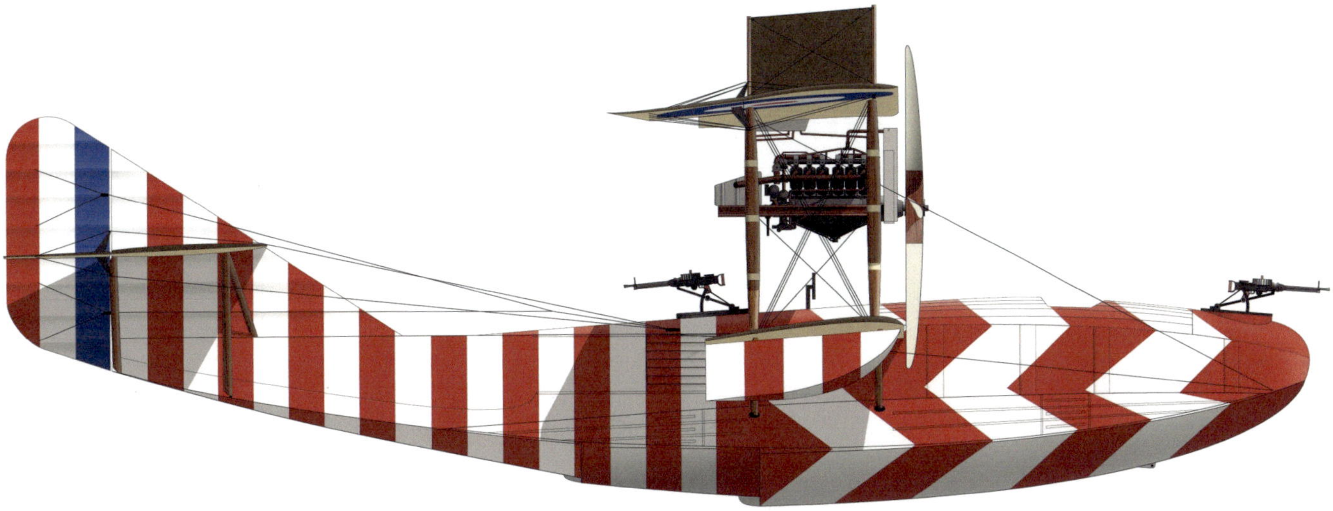

F.2A in dazzle scheme assigned to Felixstowe NAS.

F.2A N4251 in dazzle scheme assigned to Felixstowe NAS.

F.3 N4258 was delivered in July 1918 but did not get to Felixstowe until September. It survived into 1919.

Month	Patrol Days	Patrols	Hours in Air	Submarines			Zeppelins (EA)			Machines						
				Sighted	Bombed	Totals	Sighted	Attacked	Destroyed	Used	Lost at Sea	Reallocated	Fitting New Hull	Crashed	Deleted	New Machines
April 1917	12	27	108	5	3	8	—	—	—	6	8659(1)	—	—	—	—	8657 8658 86XX 8664 866X 8668
May	23	57	185	2	5	7	—	—	—	8	—	8668(8)	—	8663(13)	—	8662 8667 8682
June	21	60	239	2	3	5	—	—	1	10	—	—	—	—	—	8676 8677 8689 86XX
July	22	89	283	—	5	5	—	1	—	10	8667(2)	—	—	—	—	—
Aug	8	27	78	2	—	2	—	—	—	9	—	—	—	—	8658(15) 8682(15)	—
Sept	17	74	185	3	6	9	—	—	—	10	—	—	—	—	—	8683 8693 8694
Oct	15	56	128	3	1	4	—	—	—	10	8693(3)	8662(9) N65(10)	8678	—	—	—
Nov	16	34	66	—	1	1	—	—	—	6	8694(4)	—	8683	—	—	N64
Dec	11	30	72	—2	1	3	—	—	—	5	8676(5)	—	8677 8661	—	—	8676
Jan 1918	9	16	49	—	—	—	—	—	—	5	8683 Salvaged. (6)	—	—	—	—	8683 8677 8661
Feb	11	26	81	2	—	2	(1EA)	—	—	7	—	—	N4339	N4338(14)	—	N4338 N4359 N4282
March	19	74	258	—	—	—	(2EA)	—	—	9	—	N4282(11)	8680 XXXX	—	N64(16)	N4510 N4284 N4344 NXXXX
To 24th April	6	35	112	1	—	1	—	—	—	10	N4510(7)	N4349(12)	—	—	N64(16)	N4518 8689(17) N4345 N42XX(18) N4519 N4349
Totals	190	605	1844	22	25	47	Zeppelin EA	1	1 3	Av.8	7	—	—	2	—	

Appendix 4.1 Felixstowe War Flight (Continued)

One Year: 13 April 1917 – 24 April 1918

17 Forced landings at sea.
1 Pilot got lost.
1 Casting on Pulley Broke.
8 Engine trouble.
3 Water connections.
3 Dirty water.
1 Petrol pump failed.

The table on preceding page has been developed in the main from the charts in UKNA AIR1/6A/4/29 with corrections to illegible sections from other references where possible.

Table Notes:

(1) 29.04.17. Force landed on way to Great Yarmouth. No operational use.
(2) 6.07.17. Sank under tow.
(3) 24.10.17. Crew interned in Netherlands.
(4) 3.11.17. Towed in but damaged beyond repair.
(5) 27.12.17. Sank under tow.
(6) To H-12 Convert.
(7) 8.04.18. Sank under tow.
(8) To Killingholme.
(9) To station School.
(10) To Experimental work.
(11) To Great Yarmouth.
(12) H-12 Convert. To Great Yarmouth.
(13) 20.05.17. Crashed while landing, crew saved.
(14) Shot down 15.02.18. Crew all killed.
(15) 31.08.17.
(16) 15.05.18.
(17) H-12 Convert.
(18) Converted to F.2A.

Appendix 4.2: Typical Report Reflecting the Usual Work of Flying Boats

RAF Felixstowe 7 to 13 July 1918

Date 1918	Serial	Flight Time in Air Hours – Mins	Distance flown in Miles	Nature of Patrol	Remarks
7 July	N4359	6 – 35	403	Anti-submarine	
	N4531	6 – 35	403		
	N4538	6 – 40	403		
7 July	N4301	2 – 43	157	Hostile Aircraft patrol and Submarines	Accompanied by two land machines
	N4530	2 – 42	157		
8 July	N4539	1 – 45	125	Search for mine	Nothing to Report
	N4538	1 – 45	125		
	N4531	1 – 45	125		
9 July	N4539	2 – 25	162	Hostile Aircraft patrol	Accompanied by two land machines

F.2A N4297 was in service at Felixstowe from May 1918. Engaged with Brandenburg W.29 monoplanes on 04 July 1918 and was forced down. Recovered and repaired it was still in service at the end of January 1919.

Felixstowe Flying Boats Alightings at Sea – From 13 April 1917 to 28 August 1918								
No.	Date	Serial	Type	Cause	Landing	Returned	Lost	Remarks
1	24/04	8658	H-12	Points on magneto fused.	Good	Yes		Towed.
2	29/04	8659	H-12	Pilot lost bearings.	Bad		Yes	Crew saved.
3	03/05	8658	H-12	Casting on control pulley broke.	Good	Yes		Taxied & towed.
4	03/05	8661	H-12	Landed beside 8658.	Good	Yes		Got off & returned.
5	12/05	8662	H-12	Water connection broke.	Good	Yes		Towed.
6	24/05	8667	H-12	Water connection broke.	Good	Yes		Taxied & towed.
7	18/06	8667	H-12	Oil pressure failed.	Good	Yes		Taxied.
8	16/07	8667	H-12	Petrol failure.	Good		Yes	Lost in tow, crew saved.
9	17/07	8677	H-12	Petrol pump failure.	Good	Yes		Flown back.
10	25/07	8687	H-12	Water connection broke.		Yes		Towed.
11	21/08	8677	H-12	Reduction gear.		Yes		Towed.
12	22/08	8689	H-12	Reduction gear.		Yes		Towed.
13	22/10*	8693	H-12	Big end broke.	Good		Yes	Drifted into Holland.
14	05/11	8694	H-12	Port engine seized.	Good		Yes	Lost in tow, crew saved.
15	19/12	8689	F.2A	Throttle cable broke.	Good	Yes		Flown back.
16	19/12	8676	H-12	Petrol pipe broke.	Good	Yes		Flown back
17	27/12	8676	F.2A	Petrol pipe.	Good		Yes	Damaged by TBD, lost in tow.
18	02/01	8683	F.2A	Engine failure.	Bad	Yes		Landed downwind, beached.
19	05/02	8677	F.2A	Engine vibrating.	Good	Yes		Flown back.
20	12/02	N4349	F.2A	Magneto trouble.	Good	Yes		
21	18/03	N4282	H-12	Broken petrol pipe.	Good	Yes		Flown back to base. N4282 was Saunders built F.2A.
22	19/03	N4510	F.2A	Dirty petrol.	Good	Yes		(Landed twice).
23	21/03	8677	F.2A	Landed beside N4510.	Good	Yes		Flown back. H-12 Convert.
24	21/03	N4510	F.2A	Dirty Petrol.	Good	Yes		Flown back.
25	25/03	8661	F.2A	Water jackets cracked.	Good	Yes		Towed back.
26	08/04	N4510	F.2A	Petrol pipe broke.	Good		Yes	Lost in tow, heavy seas, crew saved.
27	14/04	N4519	F.2A	Petrol pipe broke.	Good	Yes		Flown back.
28	19/04	8689	F.2A	Engine trouble	Good	Yes		Flown back.
29	19/04	N4345	H-12	Dirty petrol.	Good	Yes		Towed into Felixstowe.
30	17/05	N4345	H-12	Landed to look at balloon.	Good	Yes		Flew home.
31	27/05	N4294	F.2A	Engine trouble.	Good	Yes		Flown back.
32	29/05	N4294	F.2A	Dirty petrol.	Good	Yes		Flown back.
33	30/05	N4344	F.2A	Short circuit.	Good	Yes		Flown back.
34	04/06	N4302	F.2A	Petrol pipe.	Good	Yes		Flown back.
35	04/06	N4533	F.2A	Broken petrol pipe.	Good		Yes	Was taxing to Dutch waters when attacked by German seaplane & set on fire. Crew saved, interned in Holland.

Felixstowe Flying Boats Alightings at Sea – From 13 April 1917 to 28 August 1918

No.	Date	Serial	Type	Cause	Landing	Returned	Lost	Remarks
36	04/06	8689	F.2A	Enemy action.	Good		Yes	Shot down, crew rescued by Dutch trawler. Interned in Holland.
37	06/06	N4345	H-12B	Enemy Action	Bad		Yes	Crashed on landing, crew rescued next day by flying boats from Felixstowe.
38	06/06	N4531	F.2A	Landed beside N4345.	Good	Yes		Took up two of crew, flown back. Note date.
39	06/06	N4530	F.2A	Landed beside N4345	Good	Yes		Took up two of crew, flown back. Note date.
40	14/06	N4298	F.2A	Landed xxx up balloon.	Good	Yes		Flown back.
41	28/06	N4540	F.2A	Landed to speak with ML.	Good	Yes		Flown back.
42	30/06	N4293	F.2A	Starboard engine seized.	Bad		Yes	Crashed on landing, crew saved.
43	04/07	N4297	F.2A	Enemy action.	Good	Yes		Towed to land. Repaired.
44	04/07	N4513	F.2A	Enemy action.	Good		Yes	Lost in tow, crew all wounded, one killed.
45	04/07	N4540	F.2A	Enemy action.	Good	Yes		Landed Yarmouth. Towed, for repair.
46	02/08	N4297	F.2A	Water jacket cracked.	Good	Yes		Towed into land.
47	02/08	N4284	F.2A	Petrol lock.	Good	Yes		Taxied to land.
48		N4543	F.2A	Engine trouble.	Good	Yes		Returned on Lighter.
2a	x/xx	8xx2	H-12	Landed to take crew off boat.	Good	Yes		Taxied & towed.
XXa	30/06	N4294	F.2A	Lighter trip.			Yes	Capsized on takeoff. Crew saved.
47a	02/08	N4299	F.2A	Locked ailerons.	Good		Yes	Also reported to have forced alighting while chasing Zeppelin with five other boats & crew interned in Holland.

* Sturtivant & Page give date as 24.10.17. AIR1/305/15/226/153, gives date as 23rd & 24th.

Notes: 1. This table has been developed in the main from the charts in UKNA AIR1/6A/4/29 with corrections to illegible sections from other references where possible.

2. The H-12 conversions are noted as F.2A boats and not H-12 Converts in the tables.

During this Period

Patrols Carried out	949
Hours in Air	3007
Miles Flown	176,333

Landings at Sea

Voluntary	9
Engine trouble & Petrol	33
Crashed getting off	1
Controls	3
Enemy Action	5
Total	**51**

Machines Returned to Base **38**

Machines Lost

In tow	5	
Crashed	4	
Interned	1	
Destroyed by crew	3	
Total		**51**

No lives were lost in any of these landings.

5. The Norman Thompson N.T.4 *Small America* Flying Boats

Above: Poor photograph showing the first N.T.4, 8338, with the original blunt bow on the slipway at Middleton. The machine has Green engines and a Davis recoilless gun mounted at the bow.

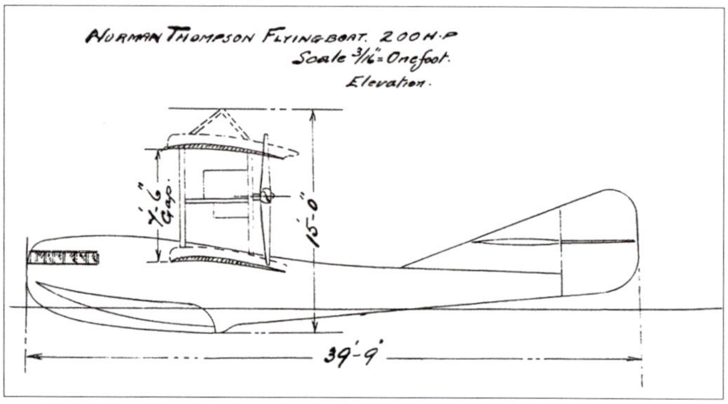

Above: The shape of the original NT hull is illustrated in this Admiralty Air Department sketch plan. The Air Department produced these dimensioned sketch plans for all machines that came through their hands.

Below: Air Department plan view of the original N.T.4.

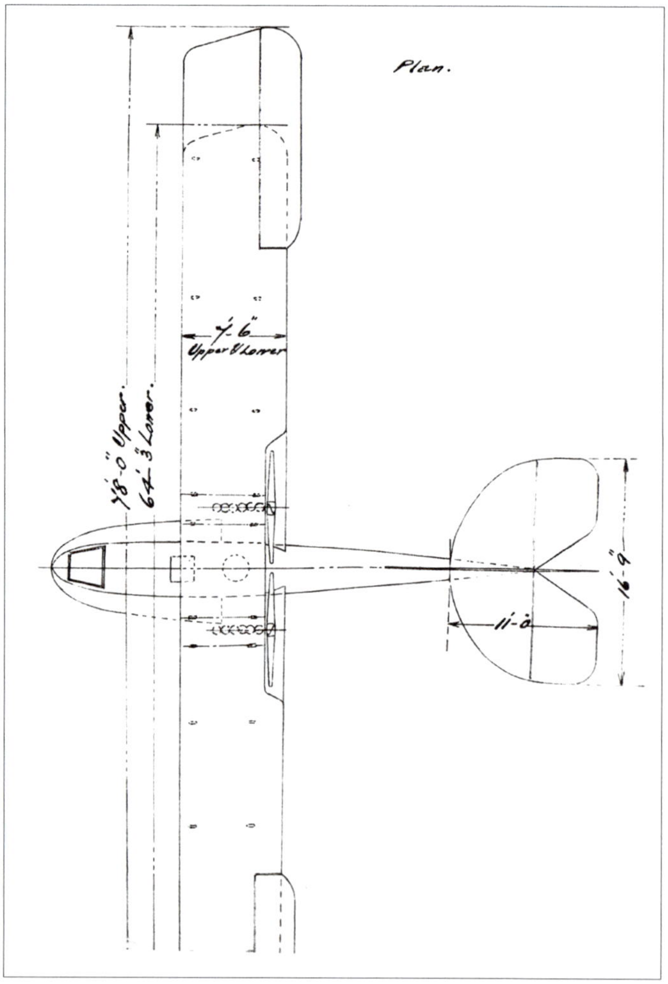

Norman Thompson's company, White and Thompson Ltd, gained the exclusive rights to sell Curtiss aeroplanes and engines in the UK in 1913. These rights were not given up and the actions of John Porte in obtaining commissions for the sale of Curtiss aircraft to the RNAS was not only contrary to regulations, but was an infringement of Thompson's rights. White and Thompson started to build a twin engined flying boat for the Daily Mail "Circuit of Britain" race. This machine was apparently based on the Curtiss flying boat that Glenn Curtiss had brought to Britain in 1913, an F-Boat with the ailerons mounted in the upper wing and not in the usual position between the wings.

The White and Thompson No.1, as the "Circuit of Britain" boat was designated, was impressed on the outbreak of war and allocated the serial 883, however no record of its use by the RNAS has been discovered to date. The company was successful in selling its N.T 2 and N.T 2a small flying boat trainers to the Admiralty. Not so successful was the company's attempt to develop a "Small America" type.

The company was renamed the Norman Thompson Flight Company in September 1915, the new company obtaining an order for six twin engined flying boats in December 1915. The new machine was also known as an "America" type although the hull owed little to Curtiss and more to Percy Beadle, the company's designer. Designated N.T.4 the machine was to be powered by two 100-hp Green straight-six engines installed as pushers. Realising that these had inadequate power two 140-hp Hispano-Suiza V-eight engines were substituted, this involving a complete redesign of the engine installation. According to testimony given at the post-war claim for compensation by

Above & Below: N.T.4 8338 with modified bow and two-pounder Davis gun installed. Note the light railway that allowed the large flying boats to be manoeuvred from slipway to hangar. Colours applied to these boats are unknown. The wing cockades are applied to the upper wing only and do not overlap the ailerons. 8338 was eventually delivered to the Calshot Flying Boat School in April 1917. Surveyed in November that year it was written off due wear and tear on the 8th of that month.

Norman Thompson, the first boat was ready for trial in May 1916, "but it was found that it would not rise from the water owing to the engine not being strong enough, and owing to the high resistance of the water occasioned by the prow being so bluff...Of course the thing was of no use and accordingly work was stopped by order of the Admiralty."[1]

The N.T.4 was a twin-engined biplane with the engines installed as pushers. The top wing was of longer span than the lower and the extensions were braced from kingposts on the upper wing. The hull of the N.T.4 was skinned diagonally with mahogany ply strips held in place by copper rivets. Calico fabric was then placed and glued to the hull and another layer of ply applied as a final layer. The fins were built separately and added after the hull was completed. A fully enclosed cockpit was provided and this was modified as the hull design progressed to the N.T.4A. Lower wing tip floats were provided as well as a bumper at the end of the hull to provide protection when on land. The machine had a single step and the rear hull was circular in section. The nose was blunt and rose vertically to the large glass enclosed cockpit and this gave the machine a pugnacious.

Post-war the Norman Thompson Flight Co Ltd made a claim for compensation due to the circumstances that arose under Contract C.P. 145936/15, dated 28 December 1915. Most of the information on the history of the N.T.4A flying boats comes from the files relating to these court actions. The

Above: N.T.4 8339 with an experimental large calibre gun, under its canvas cover, on the roof of the cockpit, the small dolly and chair horses supporting the hull front and rear while the machine is undergoing maintenance. The structure at the wing centreline is unknown but may be a W/T aerial. There is a bomb attached under the lower wing. Compare the tone of the hull against that in the photographs of 8338. This machine was fitted with bomb racks and delivered to Killingholme in January 1917. It was also at Dundee by May, and deleted on the week ending 25th May. It is not presently known if it performed active patrol duties. It was not until late 1918 that NT was constrained to use the standard 8 inch characters for their serial.

Admiralty placed the order for the supply of ten *America* type flying boats at £3,250 each, these boats were to be fitted with twin 100-hp Green engines, with propellers, instruments, etc., to be supplied by the Admiralty. Delivery was to be as early as possible. The Admiralty paid 25% in advance to enable construction to proceed.

The boat was not built to the design submitted by the Company, "the prows of the flying boats were made bluffer than in Mr Norman Thompson's design, offering a greater surface of resistance to the water. Mr Norman Thompson himself pointed out that he did not consider that the 100 horse power engine would be sufficient for the purpose that was intended, which was the purpose of lifting the boats out of the water to fly."[2] In his testimony Thompson stated that the design was "very largely" that of the Admiralty, particularly the hull design. The order for the flying boats was "in the nature of an experiment and the order was given to this Company as being the only kind of Company that could carry it out."[3] The construction was commenced until it was held up due to the modifications ordered by the Admiralty owing to the unsatisfactory operation of the Green engine. Actual work on the boats had commenced before the signing of the contract and the boats were mainly completed when the order to cease was given. They were to be stored until the necessary modifications were tried on one boat and had proved successful. The Company was told to keep its staff intact during this period. Thompson tried to obtain other contracts to keep his staff actively employed but was not allowed any such contracts according to testimony presented post-war. In June a contract for 20 FBA flying boats (N1040 – N1059) was given to the Company. These were built under the Company designation N.T.5.[4]

The work involved in order to modify the N.T.4A to take the 140-hp Hispano-Suiza engine included "disassembling wings and centre panels, redesigning, re-building and fitting eight sets of new engine mountings, strengthening trailing edge of wings in accordance with letter from Director of Air Services of June 15th to our Admiralty Inspector, Lieut-Rennie, R.N.V.R.…As regards new radiators to suit the Hispano Suiza engines, we have only ordered four at present according to instructions received, pending a final decision from the Air Department as to their suitability and size."[5] On 22 November 1916, the Company wrote to the Admiralty that "the delay arising from the unsatisfactory operation of the engines originally specified and their subsequent replacement by another of 50 per cent greater power has already exceeded eight months and pending final instructions relative to the starting of the engines and the size of the radiators, we are still unable to finish satisfactorily the first machine for acceptance tests.… There are nine (machines) in our factory in various stages of assembly, but all are ready for prompt completion when their engines are received." On 28 December the Admiralty accepted the quotation for altering "the America type Flying boats Nos 8338-8343 and 9061-9064" for the total sum of £2,216.

Above: N.T.4 8343 at Killingholme Station. This machine has black anti-fouling paint to the fins and hull bottom. This machine is also mounted on a two wheel dolly requiring support at the rear. Delivered with bomb racks for a 230-lb bomb under each lower wing, it is probable that this machine undertook anti-submarine patrols. It was placed in reserve in March and into store by January 1918, being written off the following May. Note the Sopwith Schneider floatplane in the background.

Above: The FBA flying boat was modified by Norman Thompson and given the designation N.T.5. The 100-hp Gnome Monosoupape powered N1059 was the last boat of a batch for 20 Type B boats ordered under Contract C.P.120948/16. Note the H-12 in the background.

Above: View of the cockpit of an N.T.4A.

The Company was then informed that it was to "carry out the necessary alterations to the prows of the twin engined flying boats Nos 8338-8343 and 9061-9064...in order to prevent the boats taking in too much water. It is understood that you are in possession of all the necessary particulars to enable you to carry out the alterations. To avoid delay the work should be put in hand at once."[6] The modification was in the form of a short prow added to the hull giving the boats an unfinished appearance.

The hulls were sent in February or March 1917 to Killingholme where they were to be erected on receipt of their engines. "The wing gear and flight organs had to remain stored in the Company's Works, as there was no suitable dry storage room at Killingholme."[7] This took up a lot of the floor space of the erecting shop "and we are working under considerable difficulties."[8] The machines were not complete until April 1918, when the Company had to send the stored parts and some of their employees to Killingholme to erect the machines. The post-war claim was for lost production due to having to store the hulls and aircraft parts.

The change in engines delayed the completion of the boats and it was October 1916 before the prototype, 8338, flew. Delays continued. It was reported in a meeting on 14 August 1917, that 8339 and 8340 were awaiting acceptance tests. 8339 was due to be tested that morning but the latter would not be ready for another week.[9] The NT-4 was stated to "have a rather high performance but are not very strong, the hulls having proved to be weak." It was suggested that the hulls be strengthened as this would not be a detriment in a machine meant for training. It was decided that the machines be strengthened and the fitting for carrying two 230-lb bombs be stopped. The NT-4 was to be a training machine only.[10] Sir Austin Robinson noted that the H-12 was hard to land and with its weak hull required a lot of skill. The fact that no dual control was fitted to the H-12 meant that training was carried out on the "elderly H4s, or more often in the Norman Thompson NT4s which were regarded as expendable."[11] At one time 8338 had been fitted with a Davis Gun, the Admiralty continuing to pursue the use of this recoilless gun for most of the war. 8339 had been completed by December 1916 and

Above: The hull of N.T.4A 9064 resides on the back of a lorry for this group photograph of the staff of boat builder H Williams & Co of Littlehampton. The built in lower wing roots are well displayed. 9064 was the last *Small America* of the second order for the type. This machine served at Westgate where it suffered damage to the hull and wings by inclement weather on 13 September 1917. Returned to the manufacturers for repairs it was then allocated to Killingholme, presumably as a trainer.

had been fitted to carry two 230-lb bombs and when erected at Killingholme the bomb racks were included. This was recorded as making a two hour/130 mile "S Patrol" from Dundee on 4 May 1918.[12] Capt Norman Woodward was stationed at Dundee and flew 8339 on 19 April when he spotted a submarine's swirl off Arbroath in misty weather, unfortunately they were unable to make an attack. This is probably as close as any of the N.T.4 boats came to active service. Woodward also recorded that 8340 had been modified on station with the hull fitted with large fins. Takeoff was improved but the machine was very tail heavy.[13]

A second batch of four N.T.4 boats had been ordered on 28 December 1915. These were probably originally ordered as N.T.4 machines but were delivered as N.T.4A boats. A report from the Technical Department for the week ending 18 July 1917, indicated that the N.T.4 boats were still suffering problems. It was noted that the

> *Norman Thompson Flight Company, in connection with the installation of Hispano engines in small* America *Boat, No. N.T.4A, No. 9063. This machine has been flown twice and developed different troubles on each occasion. On its first flight the engine vibrated badly and cracked the pipe connections on the radiators, owing to the way in which the radiators were suspended; instructions were given to have this altered. The second flight took place after this alteration was carried out. No further trouble was reported from this cause, but trouble arose from the exhaust up-take breaking away and fouling the propeller causing much damage. The manifolds and up-takes have been strengthened and a further test is being made on the 13th July.*

The Minute concluded by stating that a further report would be made after this test.[14] A later report mentioned that slight repairs had to be made to the hull and control column and trials would commence on 28 July. The Meeting of 28 August noted that No. 9063 was "now ready to continue tests."[15]J.D. Rennie stated that the N.T.4A had the centre of gravity in the wrong place and its correction caused "much trouble."[16]

Before the N.T.4A could enter service the Curtiss H-12 had proven to be a satisfactory flying boat, and the Felixstowe F.2A was to overcome the deficiencies of the H-12 and eventually replace it in service. There was no place for the N.T.4A even if it had proved successful. The later N.T.4A had a slightly modified hull and 200-hp Hispano-Suiza engines. The type could not be considered as a fighting machine and it is assumed that those delivered were intended for training purposes.[17] Many were delivered straight into store.

8342 and 8343 are recorded at USN NAS Killingholme. The duo was delivered to the Killingholme Seaplane School for erection on 22 March 1917. It is assumed that they were used for training purposes when in the hands of the US Navy. Both were in store by January 1918 and written off by the British the following May.[18]

An experimental boat was ordered in August 1917. This was to be a modification of the N.T.4A with the wings and tail surfaces of this boat but a new hull designed on Porte principles. It appears that at least two machines to meet the specifications were proposed. One employed two 200-hp Hispano engines but the firm did not think that this would meet the required performance and a second machine

Above: N2141 at rest on the water. The engines and cockpit are protected by canvas covers. N2141 was the second machine from the last batch of N.T.4A boats. Delivered to Calshot in November 1917, it suffered the indignity of a forced landing on the 26th, having to be towed back to port. Another forced landing due engine failure saw the machine towed but heavy seas washed the lower starboard plane away. Recorded at Portland and Cattewater in January 1918, it was finally written off in June. There is no information available for the last four machines of this batch (N2156 – N2159) and they may not have been completed.

employing three engines was proposed as an alternative. The latter proposal was not proceeded with and this may be another instance where Norman Thompson had to amend its proposals in order to obtain a contract. A note in a Technical Department Report recorded that Beadle visited the Department with "amended proposals for a twin 200 h.p. Hispano Flying Boat to Air Board specification N.2.C. The design provided insufficient water clearance for the main planes and lower tail plane, the ailerons on the lower plane were also considered unsatisfactory." Beadle was instructed to increase the height of the lower plane above the water line, increase the gap between the mainplanes; aileron on the upper planes only; and to substitute a monoplane tail in place of the present biplane tail in order to get the whole tail higher out of the water.

A later report states that the "firm is making satisfactory progress with the small twin-engined boat with reduced F.3 hull." As Norman Thompson was building Porte designed F.2A hulls they were aware of the difference from the N.T.4A's boat-built hull to those designed by Porte. Designated N.2C,[19] two machines were ordered. The resultant flying boat was powered by two pusher 200-hp Sunbeam Arab engines. The contract for two "Boat Seaplane" to the N.2.C specification was reported as "situation unchanged" in the 70th Report of the Technical Department Aircraft Production for the three weeks ending 1 January 1919, and it was recommended that the two machines be cancelled.[20]

Despite the problems, the work continued as documented in the fortnightly Reports of the Technical Department – Aircraft Production, Design Branch provide the following insights on the progress of the contract for the N.2C boats:

F/E 12.06.18. Norman Thompson Flight Co. Boat Seaplane. Two Arab. Contract for 2 machines.

1st machine: Awaiting engines. It is expected that these will be received during the next fortnight. Firm states that they can deliver this machine within 14 days from date of receipt of engines.

2nd machine: Work progressing slowly. Date of completion depends on delivery of engines.

F/E 26.06.18.

1st machine: Engines have now been received and Centre Section structure is erected. Tail and Fin are erected. Tanks are in. Work on main planes proceeding. Machine should be ready in a week.

2nd machine: Well ahead but waiting engines.

F/E 07.08.18.

1st machine: Is now complete and was taken out on August 1st for a trial. It was found that the tail section of the hull was too great. The machine rose to past the stalling angle before leaving the water. Steps are now being taken to remedy the fault.

2nd Machine: Proceeding slowly, awaiting results of No.1.

F/E 21.08.18.

An officer has visited the firm and given detailed instructions as to deepening the rear step which it is thought will mitigate the tail suction trouble. This machine is 600 lb over weight on the firm's estimate. The hull alone is 200 lb over weight.

Above: N2142 on the slipway. Note that a two wheel dolly is being used, a crewman sitting on the bow to help distribute weight so that the machine may be lifted by the groundcrew at the rear. This dolly is marked for 8339. Delivered to Calshot School in November N2142 was deleted on 2 March 1918, along with other N.T.4A boats, probably as a clerical clean up, the boat not having been in service for some time before this date.

Above: N2147 outside the hangar at Middleton judging by the light rail. The hull is now a dark colour with a light bottom. Completed in January 1918, it was at Cattewater by April. Together with N2140, N2141 and N2148, it was written off for the week ending 27 June 1918.

Above: The first N.T.4A 9061 on the water with its engines protected by their canvas covers. Delivered to the Calshot School in September 1917, it was wrecked and written off on 6 November 1917. Note how the shadow of the upper wing darkens the lower wing dope colour.

F/E 04.09.18.
1st machine, No. N82. Necessary alterations are being made to rear step to mitigate tail suction trouble.

F/E 18.09.18.
1st machine, N82. The modifications to the step have been completed and the engine bearers are now being stabilised for vibration. Shortage of material will delay completion about ten days.

F/E 02.10.18.
1st machine, N82. Work has been held up awaiting channel iron to stiffen port engine bearers. This has now been received and the machine is expected to be ready for trials this week.

F/E 16.10.18.
1st machine, N82. Has now been flown by section Test Pilot. She is improved, but poepoises badly at 30 knots. In the air she proved very tail heavy, and could not be held at alitiude with engine throttled down. A re-setting of the tail has been recommended. When this is done it is proposed to fly the machine to Grain for report and type trail. The process of further alterations at the firm's works means great delay as they are very slow.
2nd Machine, N83. All components complete and the machine is awaiting the trials of N82.

F/E 20.10.18.
1st machine, N82. This machine is awaiting trial by Section Test Pilot.
2nd Machine, N83. Waiting results of N82.

F/E 13.11.18.
N82. In abeyance, pending suggestions from firm after unsatisfactory trials.
N83. Awaiting decision made on N82.

F/E 27.11.18.
Situation remains unchanged.

F/E 11.12.18.
Situation still unchanged.

Report for period 12 to 31.12.18.
Situation unchanged. It has been recommended that these machines be deleted.

Report for Month ending 31.01.19.
Machines N82 and N83. Recommended for deletion.

The first, N82, was trialed on 1 August 1918, and proved incapable of leaving the water as tail suction caused it to exceed its stall angle. While it was being modified work was stopped on N83, the second machine that was awaiting its engines. The rear step was deepened in order to overcome the problem with take-off but the seaplane now proved to be overweight. The machine again took to the air in October and was still unsatisfactory. The Armistice ended any further prospects of a production order.

Norman Thompson were no more successful with their N.T.6 "Cruiser Flying Boat, 3 Seated Tractor, 2 – 350 R.R." that was proposed to meet the Air Department N.3b specification.[21] On 9 May 1917, it was reported as in the design stage only.[22] It was cancelled in 1917 before any work was put in hand. No further information is available for this design.

Norman Thompson was not successful in their efforts to

Above & Below: N.T.4A 9063's hull is supported at bow and tail by tressels. Was at Isle of Grain Test Depot in July 1917, where it carried out various experiments. It was moored to a buoy but the hull sprang a leak and it sank while on its way to Westgate. The machine was wrecked in towing it ashore.

Above: Detail of the engine installation and bomb rack on an N.T.4.

obtain compensation from the Government post-war, nor were their claims against the Curtiss Company upheld. The company did receive £2,250 for their N.T.2, BN.T.2A, N.T.2b, N.T.3 and N.T.4 flying boats from the Royal Commission on Awards to Inventors. This was announced in the same issue of *Flight* that noted that Colonel J.C. Porte, for improvements in flying boats, and the Curtiss Aeroplane and Motor Co, for aircraft and hydro-aircraft, each received "nil".[23]

After the war there was a deal of discussion as to the best method of building flying boat hulls. In particular the Porte and Linton Hope methods were the two that were championed by two sides in what was a long running argument. The Curtiss method of hull building was not mentioned except that Maj J.D. Rennie in reply to a remark that in Porte's time "experimental work was confined entirely to this type of

Facing Page: N.T.4A under construction at Middleton. The large radiators with starting handles in their centres are well illustrated.

N.T.4 Specifications

Dimensions	1. N.T.4 Prototype	2. N.T.4A	3. N.T.4A
Span Upper	78 ft 0 in	77 ft 10 in	78 ft 7 in
Span Lower	64 ft 3 in	64 ft 3 in	60 ft 10 in
Chord	7 ft 6 in	7 ft 6 in	7 ft 6 in
Dihedral		1° 30'	
Incidence		5°	
Length	39 ft 9 in	40 ft 9 in	41 ft 6 in
Height	15 ft 0 in	14 ft 7 in	14 ft 10 in
Wing Area		1,014 sq ft	997.8 sq ft
Weight Empty			4,572 lbs
Weight Loaded			6,469 lbs

Sources: 1. Air Department sketch. TNA AIR1/716/27/19/30.
2. Goodall, M.H. *The Norman Thompson File*, Air-Britain, UK, 1995.
3. Bruce, J.M. *British Aeroplanes 1914–1918,* Putnam, UK, 1957.

Above: The only photograph known of the N.2C. What can be seen of the boat confirms that it resembled the earlier N.T.4 design. Sunbeam Arab engines with large four-bladed propellers. The hull is hidden such that any resemblance to a Porte hull cannot be determined.

Above: An N.T.4A at rest on the water.

Above: A Norman Thompson *America* boat taxiing.

Above: An N.T.4 at an operational base in winter. This machine is supported by a substantial dolly and does not require support for the hull elsewhere.

Norman Thompson N.T.4

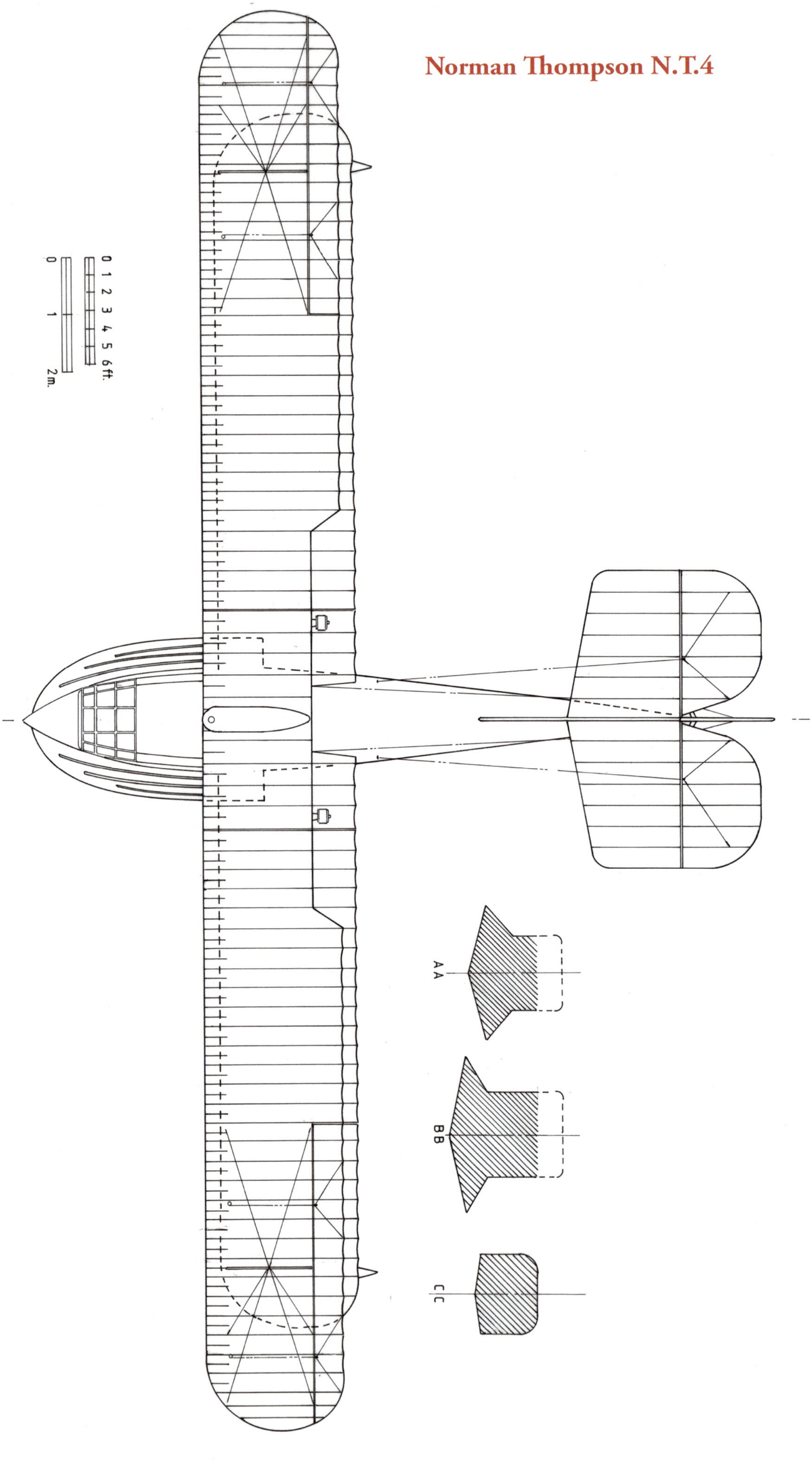

Norman Thompson N.T.4

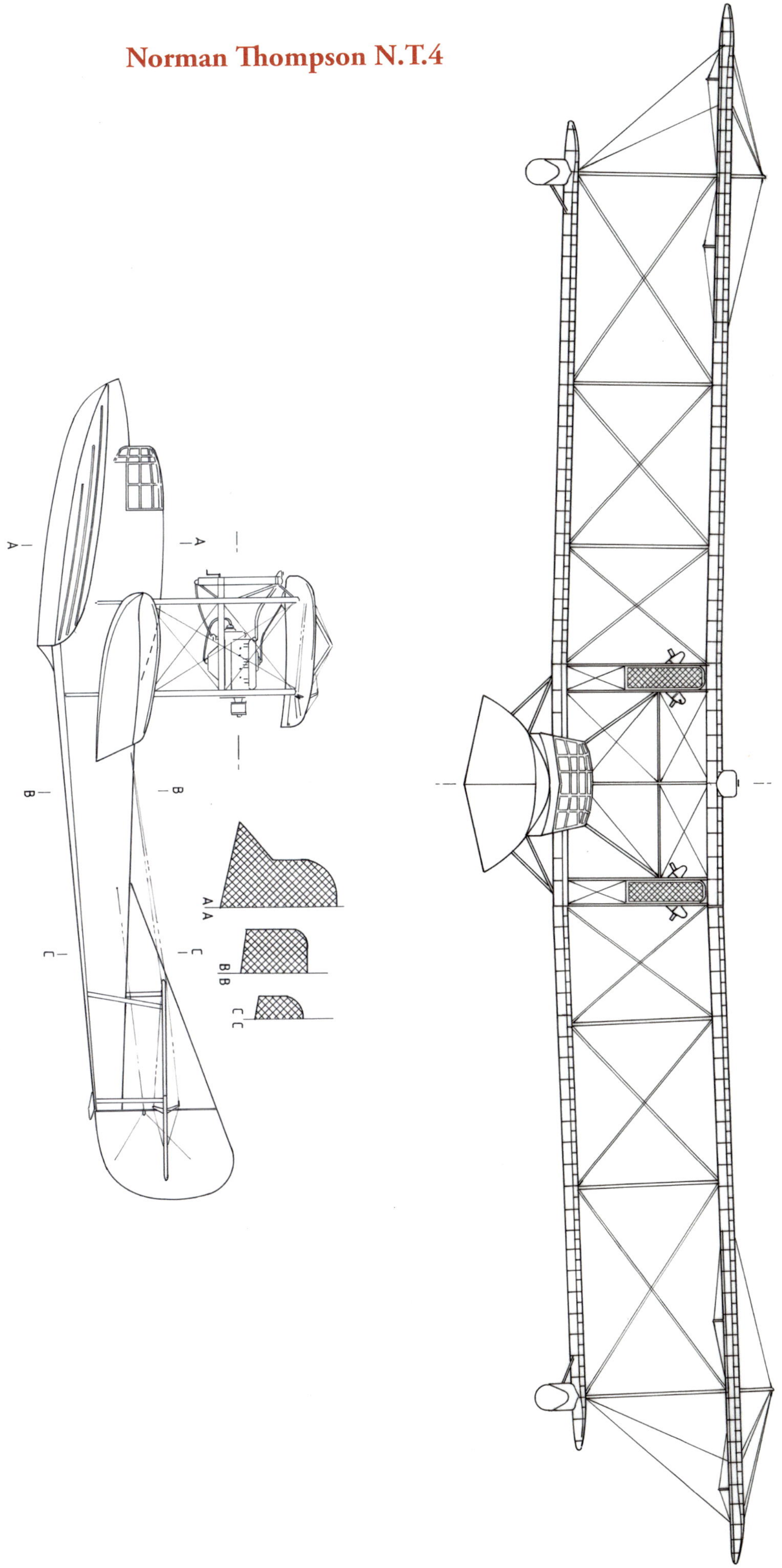

N.T.4 Serial Allocations				
Serial	**Number**	**Type**	**Contract**	**Notes**
8338 – 8343	6	N.T.4	CP145936/15	28.12.15.
9061 – 9064	4	N.T.4A *	CP145936/15	28.12.15.
N2140 – N2159	20	N.T.4A	AS12528 (BR63) 17.05.17	N2140 delivered 1 July 1918. Most to store.
N2740 – N2759	20	N.T.4A	AS32477 (BR225) 30.10.17.	Cancelled June 1918.
N18 –N19	2	N.T.6	12.01.17.	Air Cruiser. Cancelled 07.03.18.
N82 – N83	1	N.T.2C	02.04.18.	To Specification N.2C. N83 never completed.
* Sturtivant & Page (*Royal Navy Aircraft Serials and Units 1911–1919*), state that the contract was for the N.T.4A. (P.152).				

boat. This is not the case, as one firm at least, The Norman Thompson Flight Co., were concentrating on flying boats at the same time. They had little success, as their hulls were of the old Curtiss type, which had many serious defects."[24]

Chapter 5 Endnotes

1. Mr J.B. Melville, Counsel for Claimant. War Compensation Court, Norman Thompson Flight Co Ltd (in Liquidation), 22.04.1925. TNA MUN4/6435. The Air Department Report for the week ending 14.07.1916, noted the substitution of 140-hp Hispano-Suiza engines in 10 *America* Flying Boats, serials 8338/43 and 9061/64. TNA AIR1/337/15/226/247.
2. Mr J.B. Melville, Counsel for Claimant. *Ibid.*
3. The Sopwith Company had built the Bat flying boat pre-war. Norman Thompson's company probably had the most experience with building flying boats in the UK at the time.
4. According to Sturtivant & Page, (*Royal Navy Aircraft Serials and Units 1911–1919*), the hulls were built in France and Norman Thompson built the wings under Contract C.P.120948/16. Goodall (*The Norman Thompson File*) states that the FBA's hull was redesigned, the changes being such that the company called it their N.T.5.
5. Mr J.B. Melville, Counsel for Claimant quoting from Norman Thompson letter to Admiralty. *Op Cit.*
6. Mr J.B. Melville, Counsel for Claimant quoting from letter from Director of Navy Contracts of 17.01.1917. *Ibid.*
7. Minute: Mr Anderson. "Norman Thompson Flight Co Ltd, Claim £36,288.00.00. *Op. Cit.* TNA MUN4/6435.
8. Mr J.B. Melville, Counsel for Claimant quoting from letter from the Company to Air Board, 28.02.17, explaining difficulties of the Company. *Op. Cit.*
9. Minutes of the 14th Meeting of the Air Department Weekly Progress Meeting. 14.08.1917. TNA AIR1/270/15/226/113.
10. Minutes of 18th Air Department Weekly Progress Meeting. 11.09.1917. TNA AIR1/270/15/226/113.
11. Robinson, A. "The Recognition of First War Flying Boats," *Cross & Cockade Great Britain*, Journal, Vol.11 No.1 1980, P.23.
12. RAF station Dundee, Weekly Aircraft Patrol Report W/E 04.05.1918. TNA AIR1/342/15/226/271.
13. Mottram, G. "Norman Woodhead – Naval Aviator," *Cross & Cockade Great Britain* Journal, Vol.9 No.2, 1978, P.78.
14. 22nd Weekly Report of the Controller, Technical Department, for W/E 18.07.1917. Copy in RAFM J.M. Bruce Collection Box 25.
15. 28th Weekly Report of the Controller, Technical Department for W/E 29.08.1917, P.5. Copy in RAF Museum J.M. Bruce Collection Box 59.
16. Notes from interview with Maj J.D. Rennie, 15.09.1964. RAFM J.M. Bruce Collection Box 40.
17. In his correspondence with J.M. Bruce, Sir A. Robinson indicated that to his knowledge the N.T.4 never carried out a submarine patrol. To this comment Jack Bruce replied that "it would not surprise me to learn that they committed no warlike act at any time. They were most unmilitary flying boats." Correspondence with Sir A Robinson, Comments on Professor Robinson's Notes: Feb–June 1959. RAFM JM Bruce Collection Box 22.
18. Owers, C.A. "Killingholme Diary," *Cross & Cockade International*, Vol. 35 Nos. 3 & 4. 2004.
19. A "List of Contracts Since 4th August, 1914" given to Norman Thompson Flight Co Ltd refers to the N.T.2C as a "Government modification of the N.T.4A." TNA T173/469.
20. Copy in RAFM J.M. Bruce Collection Box 25.
21. There is some confusion as to the actual Specification that the Cruiser Flying Boat was designed to meet. Sturtivant & Page state it was to be built to the N.2(c) specification (P.177). This is thought to be incorrect as the Norman Thompson N.2c was designed to this specification. See Table in TNA AIR1/6A/4/47.
22. Chart. Air Board Technical Department, Experimental Seaplanes on Order, May 9th, 1917. TNA AIR1/6A/4/47.
23. "Awards for War Inventions," *Flight*, 15.01.1925.
24. "Some Notes on Flying Boats." *Journal of the Royal Aeronautical Society*. P175.

Norman Thompson N.T.4 serial 8338, the prototype. Hull shown as stained wood.

Norman Thompson N.T.4 8343 was assigned to Killingholme Station. It was placed in reserve in March 1917 and into store by January 1918, being written off the following May.

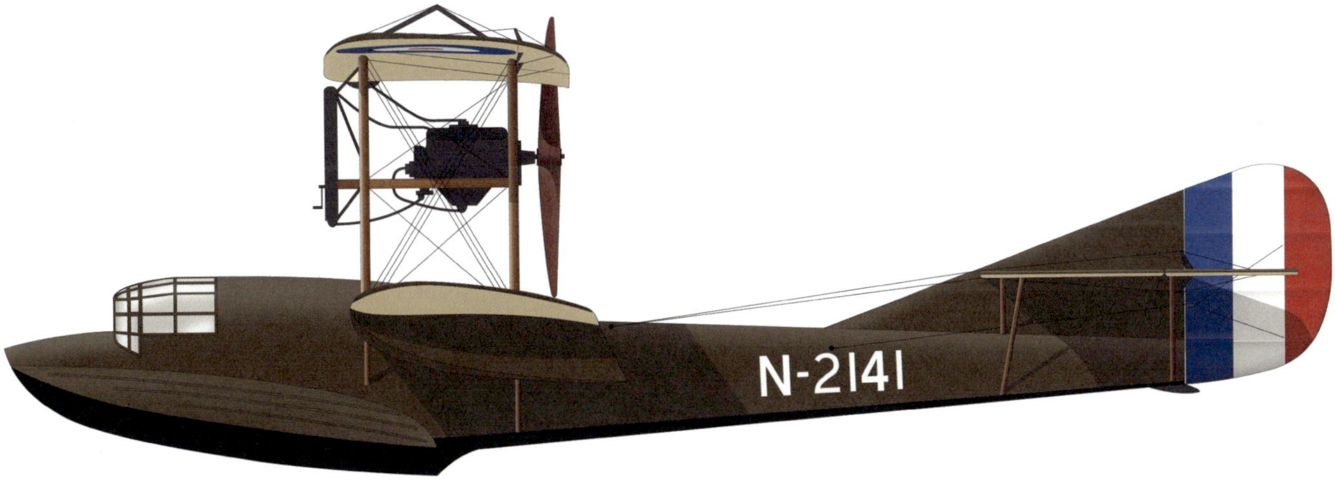

Norman Thompson N.T.4A N-2141 was delivered to Calshot 26 November 1917, and is known to have served at Cattewater later that year. It was deleted on 27 June 1918.

Appendix 1: John Porte and the Curtiss Commission Scandal

John Cyril Porte was born on February 26, 1884, at Bandon, County Cork Ireland. The son of an Irish clergyman, the vicar of Denmark Hill, he was the fourth child of seven children. Educated at private schools he joined HMS *Britannia*, the Royal Naval College at Dartmouth, at the age of 14, leaving at the end of December 1899. He then went to sea on HMS *St George* in January 1900. He remained at sea in various ships until February 1902 when, as a Sub-Lieutenant, he went to the Royal Naval College, Greenwich. In January 1904 he left Greenwich for the college at Portsmouth, leaving here in March the same year. Returning to sea in HMS *Royal Oak*, he remained at sea until Christmas 1904.

In February 1905, the same month he joined the Submarine Service, he was promoted to Lieutenant. He remained in the Service until August 1910. He served on the submarines B3 and C38. It was while in the Submarine Service that he became interested in aeronautics. Returning to sea again in August 1910 on the *Duncan*, he became sick in October 1910 having contracted pulmonary tuberculosis. Given a year's leave, he was then placed on the retired list in October 1911.

Porte was a fine naval officer and his association with the submarine service is common amongst many of the early naval aviators. After three or four months in Switzerland he returned to the United Kingdom and began a career in aviation. He had built several gliders while stationed at the Submarine Depot, Portsmouth, and also purchased a machine, a little monoplane of the Santos Dumont *Demoiselle* type.

Porte learned to fly at Reims, France, and gained French Aero Club Certificate No.548 on July 28, 1911, flying a Deperdussin monoplane. He had a successful racing season and he started an aviation syndicate with an Italian, Mr D Lawrence Santoni. This syndicate was later transferred into a company, the British Deperdussin. Aeroplane Company Ltd. The company ran a flying school at Brooklands and later at Hendon and were agents for the French Deperdussin company. Porte states that during this time he was working with aeroplanes and seaplanes.

In the latter part of 1913 he joined the firm of White and Thompson of Bognor. He worked with them for only a few months as Assistant Manager. While with this firm he first saw the Curtiss flying boat at Brighton and met Glenn Curtiss through his friend Captain Bass who had bought one of the Curtiss boats. Bass convinced Porte to join a syndicate to build/market Curtiss flying boats in the United Kingdom. Bass gained the sole agency for Curtiss flying boats and engines in Europe, and in February 1914, transferred this to the White and Thompson Co. The latter set up a flying boat school and Porte was appointed flying instructor. It appears that Curtiss was impressed by Porte's flying of the Curtiss boat.

Porte was "very struck" with the possibilities of the Curtiss F-Boat and flew it several times, including times when Curtiss was present. It was the first time Porte had seen a flying boat. It "was a boat with wings on it rather than an aeroplane with floats on it and was therefore it had much greater seaworthiness and much greater stability."[1] White and Thompson obtained the sole rights to sales of Curtiss boats in the UK.

Curtiss had brought a 1913 Model F flying boat with him to demonstrate in Europe. This boat differed from the "standard" F-Boat in that the ailerons were attached to the trailing edge of the upper wing rather than mid-wing as on most other F-Boats. The English F-Boat was detailed in *Flight* magazine of April 11, 1914. This boat was criticised as it was considered structurally weak and poorly constructed.[2]

Norman Thompson also considered the Curtiss to be deficient and set about overcoming these in his design of the White and Thompson No.2 Flying Boat. Bass crashed his Curtiss boat and it was reconstructed with wings of RAF 6 section. Fitted with a 100 hp Anzani radial engine it was markedly superior as compared with the original Curtiss wings. In this guise it was known as the Bass-Curtiss Airboat.

Porte contends that he was one of the first persons to talk seriously about flying the Atlantic as far back as 1911. He thought that the flying boat was the answer to the task of flying the Atlantic. He eventually saw Mr Tuchy of the *New York World* in London and discussed the idea with him. This scheme nearly came about but fell through at the last moment. He saw one of the directors of the English *Daily Mail* and shortly thereafter they made their famous offer of a prize for the first flight across.

Porte then obtained an introduction to Rodman Wanamaker in February 1914. Wanamaker agreed to finance the scheme using a Curtiss machine. Wanamaker's agents had already been in touch with Curtiss. Porte only stayed ten days and then returned to England.

Prior to his trip to America Porte had gone to Paris in December 1913 where he again met with Curtiss who was showing a flying boat there. Porte recalled that "Curtiss took a great liking to me and was deeply interested in my ideas for the development of his flying boat and we got on extremely well together." Curtiss and Porte discussed the Atlantic flight and Porte agreed to go to America "as long as my travelling expenses and one hundred dollars a week were paid for my living expenses." This amount was eventually paid for by Wanamaker.

It was originally suggested that the machine for the Atlantic flight be a land machine with a 200 hp motor. When he realized that such a motor was not going to materialize, Porte went back to Curtiss with the suggestion that "a large boat with 2 of his engines of 90 horse power each" would be the best machine for the proposed flight.[3] Curtiss eventually agreed and the resultant machine was an amalgam of Curtiss' invention and Porte's improvements.

In April 1914 Porte returned to America with the intention of flying back and of "helping to design the machine and prepare the boat and make all necessary arrangements for flying" the Atlantic.[4] "In fact in this particular machine the improvements were entirely my idea…this machine was built under my supervision and I also made a number of suggestions with regard to the alterations to the hull which were carried out and my suggestions were taken up by Mr. Curtiss."[5]

Porte had worked closely with Norman Thompson on the work of improving the F-Boats. The experience was his introduction to his practical knowledge of flying boat construction, and he took this expertise with him to America and it enabled him to carry out the work of designing the *America*.[6] Casey gives credit for the design of the *America* boat to B Douglas Thomas, an Englishman that Curtiss met on his European trip in 1913. Thomas designed the Curtiss *Jenny* and it would be logical that he would have had a great input into the design of the boat, especially the wing structure.

Wanamaker wanted a mixed American and British crew as he wanted the flight to celebratev100 years of peace between the USA and Britain. Curtiss was so impressed with Porte that he had invited him to join the crew of the proposed trans-Atlantic boat as pilot. Lt John H *Jack* Towers, USN, was suggested by Curtiss as the American half of the crew. Towers was a pilot as well as an expert navigator. Towers had flown the Navy's first hydroaeroplane, the Curtiss A-1, and had carried out experimental work with Curtiss on Naval aircraft. The USN had difficulty in accepting that a serving officer would be co-pilot to a British citizen, and the American crewmember was not resolved when Towers was ordered back to duty, sailing for Mexico in April 1914.[7]

The outbreak of war saw Porte pack up in a hurry and leave America. He presented himself to the Admiralty and was appointed command of Hendon RNAS Training School.

In late1917 Porte, William Agustus Casson and Lyman J Seely were the subject of criminal proceedings for contravention of the provisions of the Prevention of Corruption Act 1906, in respect of commissions they received between January 1915 and about June 1916 from the Curtiss Aeroplane Co of New York in connection with aeroplanes, flying boats, and motors, etc., supplied by the Curtiss Company to the Admiralty between August 1914 and June 1916.

At the time Porte was a Wing Commander in the RNAS, Casson was a barrister and an intimate friend of Porte, while Seely, an American subject, was the agent of the Curtiss Company in London.

Casson had connections with various aircraft companies for about two years before the war and was an intimate friend of Porte. When Lt Porte was invalided out of the Submarine Service in October 1911 with tuberculosis, he became a director of the British Deperdussin Company. When that company failed at the end of 1913, Casson was appointed Receiver and Liquidator. Porte then joined White and Thompson Ltd, seaplane manufacturers of Bognor in September 1913. This firm had an agreement with the Curtiss Company for the agency rights to sell Curtiss flying boats in the UK. Porte remained with the firm until April 1914.

When Porte was selected as the pilot of the proposed trans-Atlantic flight by the Curtiss *America* flying boat, a verbal agreement was made between Casson and Porte that the former would look after the interests of the project in Europe and make any disbursements necessary on this side of the Atlantic on condition that any profits coming to Porte would be shared equally between them. Porte claimed that Glenn Curtiss had given an undertaking in a letter that in consideration of his contribution to perfecting the design of the *America* flying boat, he would receive a 25% commission on the selling price of all flying boats for which orders might be obtained by him or through his instrumentality. The actual letter was stated to have been lost when Porte left the USA in August 1914.[8]

With the outbreak of war, the flight was cancelled and Porte left for England. He offered his services to the RNAS and was appointed a Squadron Commander in the RNAS and had remained an officer of the RN ever since. In August 1914 he discussed with Commodore Sueter the purchase by the Admiralty of two flying boats *which he had designed (author's italics)* from the Curtiss Company, and at the request of Commodore Murray Seuter he negotiated the purchase of these two *America* boats. The initial prices asked for by the Curtiss Company were considered too high and Porte negotiated lower prices for the purchase and this was approved by Seuter and sanctioned by the Board and on 29 August the Admiralty cabled the Curtiss Company an offer to purchase that was accepted.

In September and October1914, again at Seuter's request, Porte negotiated the purchase of engines and spares for these boats and inspected them on delivery. Porte had suggested the purchase as he had first hand knowledge of these boats. In November 1914, he was again requested to negotiate the purchase of four more flying boats.

Porte regularly and constantly reported for the Admiralty on the flying boats purchased from the Curtiss Company as he was "admittedly an expert on the subject."[9] While it was admitted that Porte had never revealed his financial interest in the flying boats built and supplied by the Curtiss Company to any officials at the Admiralty, there was "nothing to show either in his discussions with Commodore Sueter or in his reports, that he allowed his opinions or advice as to the merits of the flying boats to be in any way coloured by his financial interest. The evidence before us shows, that the flying boats, for which he was responsible, have turned out a success."[10]

Sometime in August or early September, Porte informed Casson of the alleged arrangement with Glenn Curtiss. Casson pointed out to Porte that he should have his "proper share" of the profits. Porte was well aware that by reason of his position as a serving officer, he could not have business relations with the Curtiss Company. Accordingly, on 10 September 1914,

Porte wrote to Casson that as

> *The Admiralty having given me an appointment I must give up all my business connections but as these may be of some value I don't want simply to drop them and I am willing to act on your suggestion and transfer everything to you to do your best with. Our partnership must come to an end so far as I am concerned and I therefore hereby transfer, assign and makeover absolutely to you all my interests in the Atlantic Flight, in the Curtiss Flying boat and in my Contacts with Mr Glenn Curtiss and the New York World and any others. The consideration for the foregoing assignment as between us is that you undertake to discharge all debts owing by me in connection with the Atlantic flight and to proceed to America at your own expense if the business at any time demands it.*[11]

No mention of the ¾ to ¼ distribution of profits was made in the document as this was verbal between Porte and Casson. Later arrangements were made by which the Curtiss Company paid 15% commission on ALL Admiralty orders and this was split between the three principles to the fraud. Porte must have known that his former employer, the White and Thompson Company, had the exclusive rights to the sale of Curtiss flying boats and motors in the UK, so what was the value of his contact with Curtiss?

The basis of the Treasury's case was that the Curtiss Company was paying a commission of 16% on orders it received from the Admiralty.1% of this was considered to be Seely's legitimate commission. The remaining 15% was divided equally between Seely and Casson. However, Casson's 7½% was then divided with Porte receiving ¾ and Casson ¼. By arrangement with the company, the commission was paid to Seely who then paid over to Casson his half share. On 12 February 1917, Casson was to cross the Channel and fearing that if he were drowned then Porte's interest might be jeopardised, sent home a signed document wherein he recorded Porte's rights to ¾ of the commissions received by Casson.

A Curtiss Contracts Committee, presided over by Mr JG Butcher, KC, MP, was set up to examine the contracts between the Admiralty and the Curtiss Company.

The *America* boats were delivered towards the end of October 1914, and after Seuter and Porte had a trial flight in her on 29 October, Seuter reported to the Board the next day that the type

> *had a great promise, whereupon the First Lord on 31st October, 1914, gave direction to "order a dozen as soon as possible." Eight boats were accordingly ordered from the Aircraft Company in this Country and four were ordered from the Curtiss Company.*[12]

Porte was requested to obtain prices for these latter four boats and 20 Curtiss OX motors. On 7 November, Seuter reported to the Director of Contracts that the prices were reasonable and recommended acceptance of the Company's offer. However, Porte informed Casson on the same day what had occurred and on the 7th, three days before the Admiralty's cable, Casson despatched a cable in the name of Porte to Glenn Curtiss as follows:

> *Proceed with four boats immediately order will follow William Casson Essex Court Temple who was our agent for Atlantic Flight is acting for us in all matters relating to your business in this Country. Please deal direct with him.*[13]

The letter confirming the Admiralty's cable was signed and "would in the ordinary course have been posted on the 11th to the Curtiss Company direct" was handed over to Mr LJ Seely, who had arrived in London on November 10.

Seely was at one time a clerk in a motor business and in July 1913 was appointed Publicity and Sales Manager for the Curtiss Company with a small salary and a 1% commission on sales. He had visited the UK during 1913 or early 1914, during which visit he continued to receive his salary and commission plus expenses. The Curtiss Company was then and at the outbreak of the war, a small concern. According to the evidence given to the Butcher's Committee by Mr Macdonald, the Vice President of the company, the Curtiss Company was a small company "just built up round a barn and round Mr. Curtiss development work and they did not have any very strict business methods and an order for $150,000 worth of aeroplanes would probably look to them bigger than all the capital of the Bank of England. They were all farmer boys; it was just like going into small English village and picking up a kite and that kite developing into an aeroplane and that developing into an Order." The Curtiss Company had not obtained any orders from the Admiralty before the war and Seely "was wholly unknown to the Admiralty until he was introduced there by Commander Porte."[14]

Seely arrived in England on 10 November 1914. Casson and Porte knew of his impending arrival. Seely and Casson met at the Savoy Hotel and the question of Commissions was discussed between them. The accounts given by Casson and Seely of this meeting differ in detail. Both agreed that the agreement re commissions was not agreed to at their first meeting, but they eventually agreed to divide all commissions that the Curtiss Company would pay on all contracts made with the Admiralty between the two equally. For the scheme to succeed, orders had to pass through Seely and not be sent directly to the Curtiss Company.

Steps therefore had to be taken to ensure that the letter of the Admiralty of 11 November for the Curtiss boats did not go directly to the Curtiss Company but to Seely. Porte then took Seely to the Admiralty and introduced him to Sueter as the representative of the Curtiss Company. Sueter told the Committee that to the best of his recollections, the first time he saw Seely was on 14 November 1914, in connection with the purchase of some Curtiss R2 aeroplanes. The Committee believed that by some means Porte had the letter of confirmation of the order for four flying boats handed over to Seely.

A letter to Seely from Cassons was found in the possession of the Curtiss Company in New York during the enquiry, and provided the details of this transaction. Cassons wrote:

I find that it has been possible to stop the order for the other four boats, as I got a message to that effect when I got back here.

I have arranged with Porte to call at the Savoy for you tomorrow morning at ten o'clock sharp, and he will take you to the Admiralty and get the formal order handed over to you.

Casson asked Seely to acknowledged the receipt of this letter, which he did at the foot of this letter and wrote that he

hereby acknowledge that I have received an order from the British Admiralty for four Curtiss Flying Boats which was obtained through your efforts and I agree that this order is accepted on the basis of commission under the arrangements now in existence or hereafter made by or on behalf of Mr Glenn Curtiss.[15]

Seely then cabled Mr Genung, then vice-president of the Curtiss Company,

Shall I arrange to give Casson 15 percent if able to compromise with White and Thompson. Good business in prospect.

On 14 November Genung approved "if discount does not exceed 15% maximum. When prices quoted did not anticipate commissions."[16]

Despite rigorous attempts to discover who handed the Admiralty's letter of 11 November to Seely, the Committee's enquiries proved fruitless.

The orders from the Curtiss Company were of small account being only for seaplanes, engines and spares, but were soon to extend to large orders for aeroplanes.

Commander Spencer Grey had been selected to command a squadron for service at Dunkirk but was detained in the UK by the lack of suitable aeroplanes. He had made the acquaintance of Seely at the Royal Aero Club and on hearing good reports of the Curtiss aeroplanes, namely the R.2 with 160 hp Curtiss engine and the J.N.2 with 90 hp Curtiss engine, the Admiralty ordered substantial numbers of these aircraft. Grey went to America with Seely in December 1914 to superintend the production of the JN machines, returning with Seely and the first four aircraft in February 1915.

In March and April 1915, before any exhaustive trials had been carried out on these machines, very large orders were placed for aeroplanes, seaplanes, engines and spares. The Committee examined these orders in detail as "there is evidence that Squadron Commander Spencer Grey's name has been mentioned in America in connection with the Commissions." The manner in which the large contracts were placed for untried machines suggested that the Air Department had been influenced by advice that was not disinterested in regard to the matter. The Committee concluded that "we are satisfied that there is no ground for the rumours that circulated in America or elsewhere with regard to" Spencer Grey's involvement.[17]

The total value of the orders with the Curtiss Company was in the order of $11,000,000. The aircraft were all to be delivered by September 1915. The total value of material delivered by September was about $5,000,000 for aeroplanes and spares, and about $3,000,000 for seaplanes and spares. There was no evidence that the Curtiss Company's executives knew that Seely was receiving half the commissions. Casson had, by letter of 15 December, authorised Seely to receive the commissions on his behalf, thereby concealing Seely's own share of the commissions.

Matters came to a head by June 1916 when deliveries were very late. Lt Neilson, RNVR, was sent to America in August 1916, to report on the position of the contracts with the Curtiss Company. Nielson reported that a commission of 15% was being paid by the Company on all orders from the Admiralty. Seely went to see Commodore Seuter and the matter of commissions was discussed. Seely said that as an agent of the Curtiss Company he was legally entitled to a commission. To Seuter's question of whether any officer in the Air Department had taken a commission, Seely answered "No, not a single one."[18] A general revision of contract arrangements was undertaken, and at the request of the Admiralty, the Company discontinuing its payment of any further commissions on Admiralty orders.

At the trial of Porte and Casson, evidence was presented that the Admiralty's representative in America, Capt Elder, knew of the commissions as the Admiralty's new contract with the Curtiss Company inserted a clause prohibiting the payment of commissions, and both Seely and Casson had appealed against this prohibitive clause, and Elder "declined absolutely" to do so.[19]

Mr Butcher's Committee completed its investigations by late May 1917.[20]

It was the opinion of the Treasury that the arrangement where by Porte assigned to Casson all his interests in Curtiss flying boats and contracts in his 10 September 1914, document, was a sham. The payments to Casson were in reality payments to Porte and this was borne out by the facts of how the commission was split up. This was in breach of Porte's "duty as an Officer of the Royal Navy and to the detriment and in fraud of the Naval Service of the Crown." The opinion went on to state that a "judgement against Porte for the full amount for money had and received" estimated to be £130,000, "would not touch the £16,000 of which Casson has possessed himself and I think it is well worth while to take proceedings to which Casson would be a party."[21]

It was held that both Porte and Casson were aware of the impropriety of Porte accepting a commission on goods ordered by the Admiralty from the Curtiss Company. For the purpose of carrying out their arrangement they required a third person who would be acceptable to the Admiralty as the representative of the Curtiss Company. They found this person in the person of Seely.

In the early part of 1915 monies were being received by Casson with respect to orders for aeroplanes with which Porte had no concern and which he had no interest such as he alleges he had in the case of the flying boats. It was alleged that Casson had purchased motor cars for Porte and had paid off a debt of Portes out of these funds.

Casson admitted to receiving £64,797.10.7, of which ¾ went to Porte.[22]

Flight reported that Casson and Porte attended Bow Street Police Court before the Magistrate, Sir John Dickinson, on 3 August 1917, in answer to a summons that they had contravened the Prevention of Corruption Act 1906. Casson was leaving for France when he was arrested. It was not suggested that he was absconding, although he had a letter on his person from Porte that contained the passage that "Things are going to move, and I think it most desirable that you should be off as soon as possible." Porte had to leave the court as he was taken ill, collapsing while giving evidence, the effect of his continuing tuberculosis.

At the next session of the Court on the 10th, it was revealed that Porte could not *attend as he was suffering from an advanced stage of tuberculosis "and ought to have some weeks rest to avert some dangers which threatened him." The committal hearing continued in his absence.*

Porte did not appear at the next hearing due to health reasons. Seely was not present and it was believed he was in America. Casson was committed for trial at the Central Criminal Court while the summons on Porte was adjourned *sine die* due to his health.[23]

Although Commodore Seuter had initiated the Court proceedings, in his evidence he stated that the proposal to purchase the Curtiss boats came from him and not from Porte. Both he and the head of the Contracts Department used Porte to negotiate the contract as Porte was well known to the Curtiss Company.[24]

The Attorney-General announced at the opening of the trial that Porte wished to enter a *nolle prosequi.* He explained the reasons for withdrawing the case against Porte as follows.

> *At the outbreak of war Commander Porte was in America occupying a commercial position in the aeronautical world which was a very advantageous one. Immediately on the outbreak of war he abandoned his position, came back to England, and placed his services unreservedly at the disposal of his country. At that time and now he was suffering from a most grave haemorrhage of the lung. At the present Commander Porte is doing invaluable work at the Admiralty in regard to the national defence, and the Admiralty were most anxious to retain his services. The progress of the malady from which he suffered was such that it was not possible to suppose that in any event the period for which his services would be at the disposal of his country would be a very protracted one. All the money paid to Porte, with the exception of £10,000, which has been disposed of, remained in his possession and the balance would be handed over by his representatives to the authorities. When Casson's house was searched by the police they found securities for the commissions paid him to the value of £16,000, and these are now in the possession of the authorities.*[25]

At the Central Criminal Court on November 19, Casson pleaded guilty to 12 counts of an indictment charging him with giving a gift to Wing-Commander Porte, an agent of the Crown, as an inducement for showing favour to the Curtiss Aeroplane Company. He was fined £500 for each of the 12 counts. The Crown was satisfied that Casson had no resources to enable him to discharge the fine, but Casson was willing to pay the £6,000 out of the moneys the police held, and he would restore the balance of the money to the authorities.

It was rumoured that the charges had been brought in order to disgrace another senior officer but Porte was caught up in the scandal. The judgement allowed Porte to go free on his payment to the Crown of several thousands of pounds that had been paid into his bank account. His wife was allowed to keep the funds that had been paid to her. His legal representative, Casson, was forced to pay all the commissions he had received due to Porte under the pre-war agreement. This judgement was criticised by CG Grey of the *Aeroplane* as the equivalent of allowing a burglar to go free on condition he return the majority of his loot. If innocent Porte should have been allowed to keep the funds that were due to him. What his superiors thought of Porte is revealed by the fact that immediately on his release from arrest after pleading guilty to a grave charge; he was not only reinstated at Felixstowe but given command of the war station as well. He was made a Companion of the Order of St Michael and St George, and made a full Colonel in the RAF.[26]

Porte remained in charge at Felixstowe after the Armistice. He then devoted himself to the development of large civilian flying boats. Every prospect of success seemed to be his. He had gathered a staff of flying boat experts and had adequate finance, and had interested one of the country's best manufacturers of aircraft in building his designs. Then he suffered another attack and died.

Flight in its December 25, 1919, issue, previewed the Gosport Aircraft Company's proposals for the year 1920. The Solent based company had manufactured F.5 flying boats during the war. The flying boats shown in the drawings accompanying the article showed the work of Porte and were based on the types that "had proved so successful during the late War (*and*) were considerably improved and modified to suit commercial requirements." The lineage of the Felixstowe F.5 and the triplane Felixstowe Fury could be seen in the company's proposals. Porte's death on Wednesday 22 October 1919, at Brighton, spelt the end for the Gosport Aviation Company. Without his leadership and zeal there was no one to carry on the proposed plans for large commercial flying boats. His remains were buried in West Norwood Cemetery.

Who then was John Porte? That he was admired for his courage is evident in the respect in which he was held by the men under his command. Aircrew appreciated that he listened to them, flew with them in combat and tried to solve their problems. Professor Sir Austin Robinson was a Felixstowe boat pilot and summed up the feelings of those wartime pilots who flew with Porte or in his designs. Robinson thought that if Porte had not volunteered to fight but had remained a civilian to design and develop flying boats he would have been compensated the same as Robert Blackburn or Thomas Sopwith. "Porte's crime was that he was not content to sit back and look on, but wanted to share in the fighting."[27]

He was not an aeronautical engineer but a practical man who learnt by trial and error. There is no doubt that Porte and Casson knew that they were engaged in an illegal activity. The fact that the letter for the first contract with Curtiss went to Seely, rather than being sent to the Curtiss Company in the USA, must have been due to Porte's influence. Once Seely was in the ring, then the commissions were paid by the Curtiss Company in a manner that covered Porte's involvement completely. While Porte was forthright with his admissions to the Committee, no warnings were given him that his answers could be used against him in any criminal action that might arise from the Committee's investigations. That the charges should be brought and publicised in the middle of a war, and then the participants be let off with what was in effect a slap on the wrist, shows poor political judgement on the part of someone. It would be interesting to learn of the German reaction to the reporting of the trial.

The author knows of no papers left by Porte, so his motivations must remain a mystery. What is evident from the correspondence and evidence obtained by the Commission is that Casson always acted scrupulously in preserving Porte's interests. When he thought he might be killed in his travelling to France, he drafted a new will in which he set out exactly what he had done with Porte's funds so that Porte would not suffer a loss if he was drowned. He wrote that his executor

> *Must hold certain of these* (securities and moneys in my safe) *in trust for Comm J.C.Porte...Porte is the real owner of three fourths and I own one fourth and I hold everything in trust to be divided in this way.*[28]

Appendix 1 Endnotes

1. Statement of Commander John Cyril Porte of 1st August, 1917. P.3. Copy in The Royal Commission of Awards to Inventors, Claim No.580. Claim of Mrs Mynne Porte. Copy supplied by late Ray Sturteviant from original in UK National Archives.
2. Other sources state Porte was so impressed that he convinced his employer, White and Thompson Ltd to purchase the boat and the rights to build Curtiss designs. The company became the Norman Thompson Flight Co on 4 October 1915.
3. Statement of Commander John Cyril Porte of 1st August, 1917. P.10.
4. Statement of Commander John Cyril Porte of 1st August, 1917. P.5.
5. Statement of Commander John Cyril Porte of 1st August, 1917. P.10/11.
6. Goodal, MH, *The Norman Thompson File*, Air Britain, UK, 1995, P.77
7. George EA Hallett was selected as the co-pilot after Towers departure. Casey, LS. *Curtiss- The Hammondsport Era 1907–1915*, Crown Publishers, NY, 1981, P.191.
8. It appears that no one person designed the *America* flying boat. The aircraft as first flown was not capable of flying the Atlantic and much work was carried out with modifications to the hull and at one time a third engine was installed on the top wing. The hull had "fins" or as they would later be termed, sponsons, added to improve take-off. Typical of Glenn Curtiss' operations, much was trial and error.
9. *Curtiss Aeroplane Company. Payment of Commission on Admiralty Orders placed with the Company - Criminal Proceedings against Squadron Commander J. C. Porte R.N. and others – Arrangement attending of witnesses production of evidence – Photographic Copies of Documents Connected with the case. Information obtained from certain banks by Order made Defence of Realm Regulations.* Mr J.G. Butcher's Committee's Report, Memorandum on Commander Porte's Position. Hereunder referred to as Case 838. Original in UK National Archives.
10. Mr Butcher's Committee's Report, Memorandum on Commander Porte's Position. Case 838.
11. Copy in Mr Butcher's Committee's Report. Case 838.
12. *Ibid.*
13. *Ibid.*
14. Mr Butcher's Committee's Report, P.11. Case 838.
15. Copy printed in Mr Butcher's Committee's Report, P.15. Casson's letter is dated 11.11.1914 while Seely's acknowledgement is dated the 12th. Case 838.
16. Copy printed in Mr Butcher's Committee's Report, P.16. Case 838.
17. Case 838.
18. Evidence at committal trial of Porte and Casson. "Legal Intelligence", *Flight*, 16.08.1917, P.850.
19. Evidence at committal trial of Porte and Casson. Reported in "Legal Intelligence", *Flight*, 23.08.1917, P.876.
20. Letter to First Lord of the Admiralty, 28.05.1917. Case 838.
21. RE:Porte Casson & Another. Opinion of Mr G.W. Ricketts, 14.09.1917. On 27.10.1917, Ricketts put forward the opinion that there was little chance of obtaining the return of the monies. He considered that after the result of the criminal trial then an action for conspiracy could be launched against Porte and Casson "Seely is inaccessible and his share in the plunder is out of reach of British Courts." Case 838.
22. Mr Butcher's Committee's Report, Amount of Commissions Paid, Case 838.
23. "Legal Intelligence", *Flight*, 23.08.1917, P.876.
24. Penrose, H. *British Aviation – The Great War and Armistice*, Putnam, UK, 1969, P.327.
25. "Legal Intelligence", *Flight*, 22.11.1917, P.1240.
26. "Honours", *Flight*, 06.06.1918, P.608. The award was announced on 3 June.
27. Goodall, MH. *The Norman Thompson File*, Air Britain, UK, 1995. P.76.
28. Letter from Casson of 12.02.1917. Copy in Case 838 file.

N4082 was one of a number of F.2A boats that were used by the USN until their H-16 boats were brought up to patrol standard. The US cockade is unusual as the USN did not apply national markings to the hull.

F.2A N4512 in dazzle scheme assigned to Great Yarmouth NAS.

Late Curtiss H-12 from Calshot Naval Air Station. Two large "C" were applied to the upper surface of the top wing.

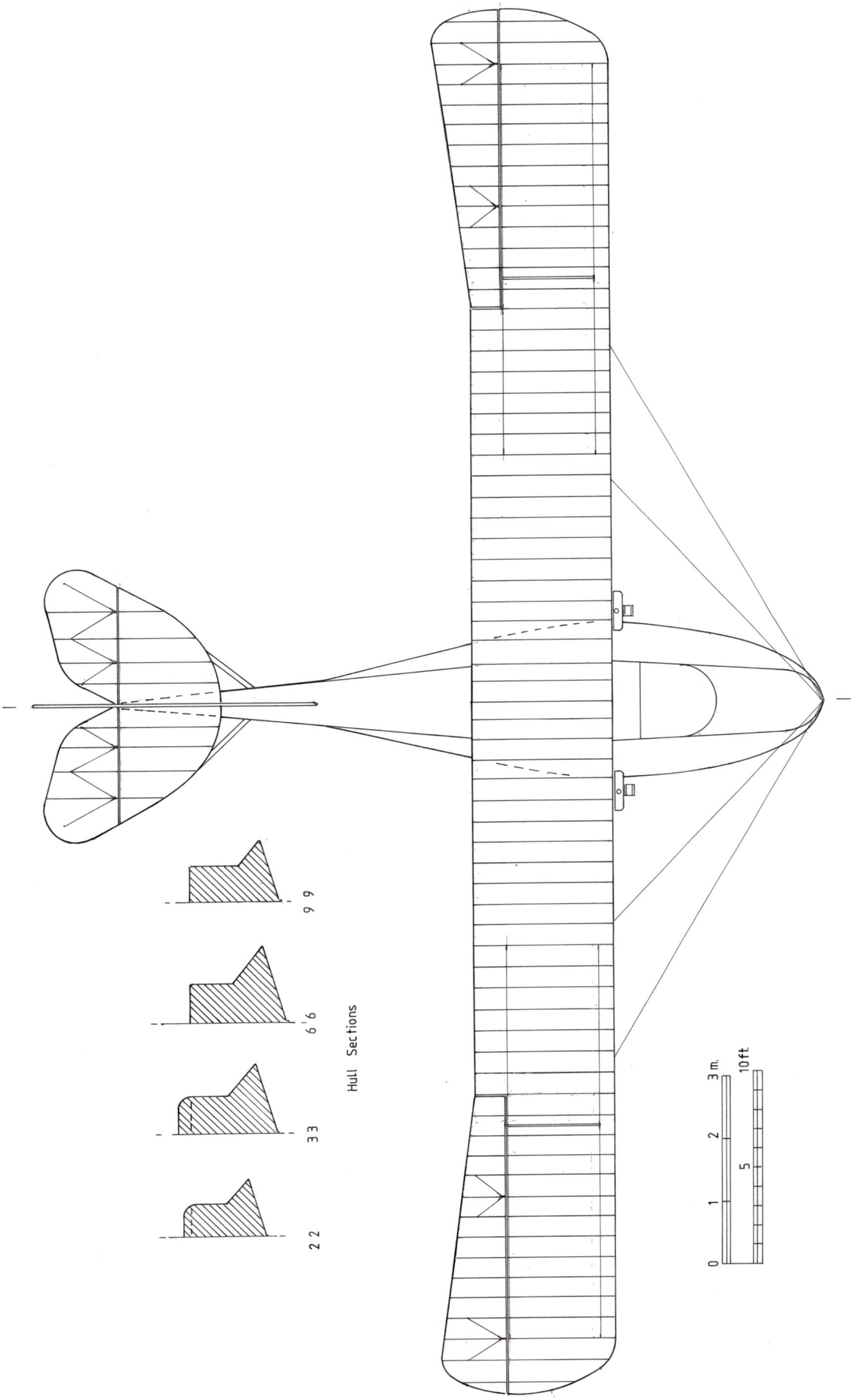

Hull Sections
2 2
3 3
6 6
9 9
0
1
2
3 m.
5
10 ft.

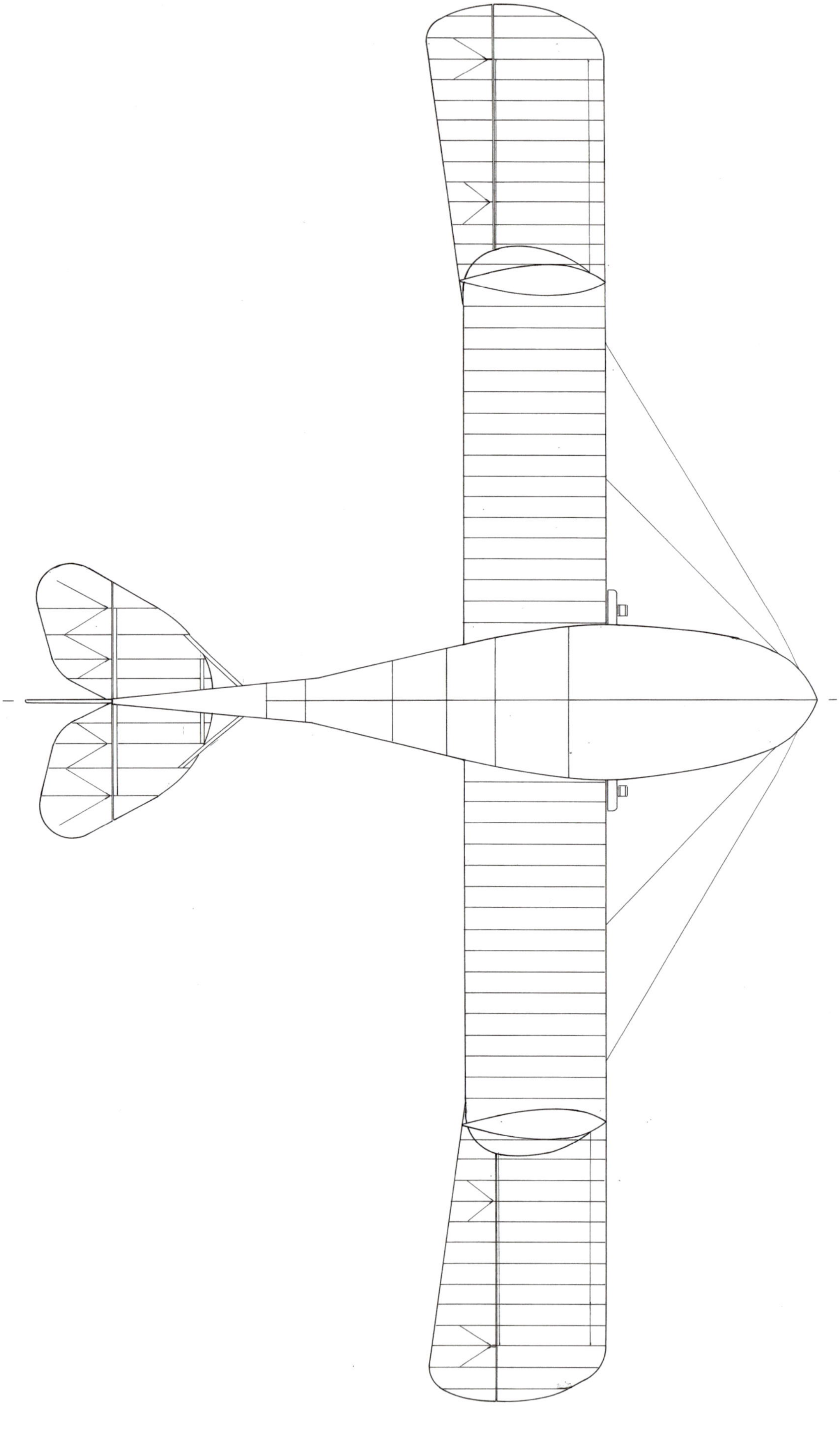

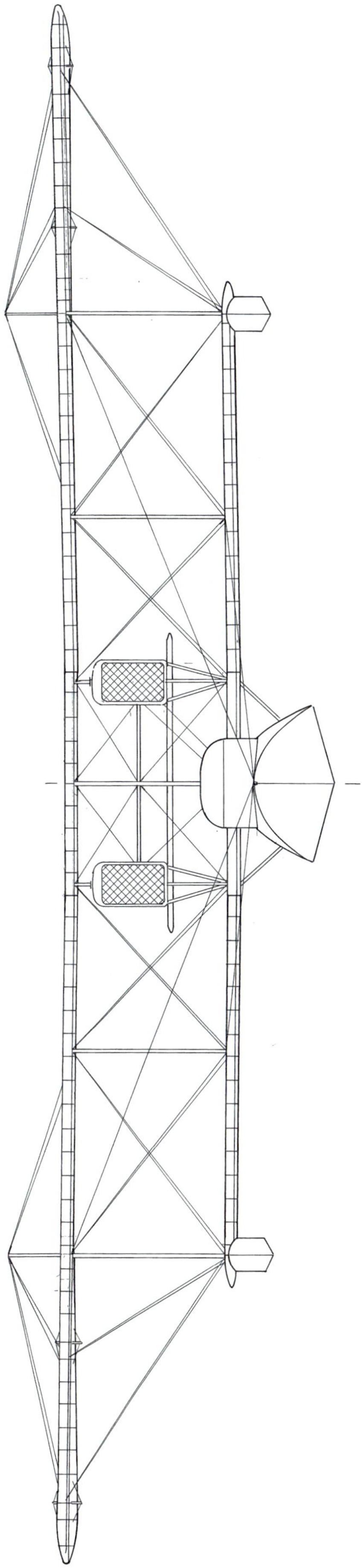

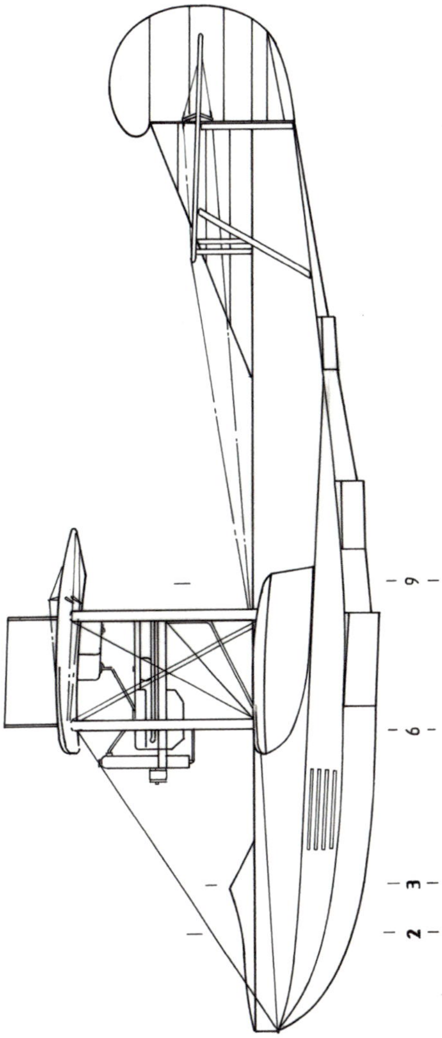
9
6
3
2

Felixstowe F.3

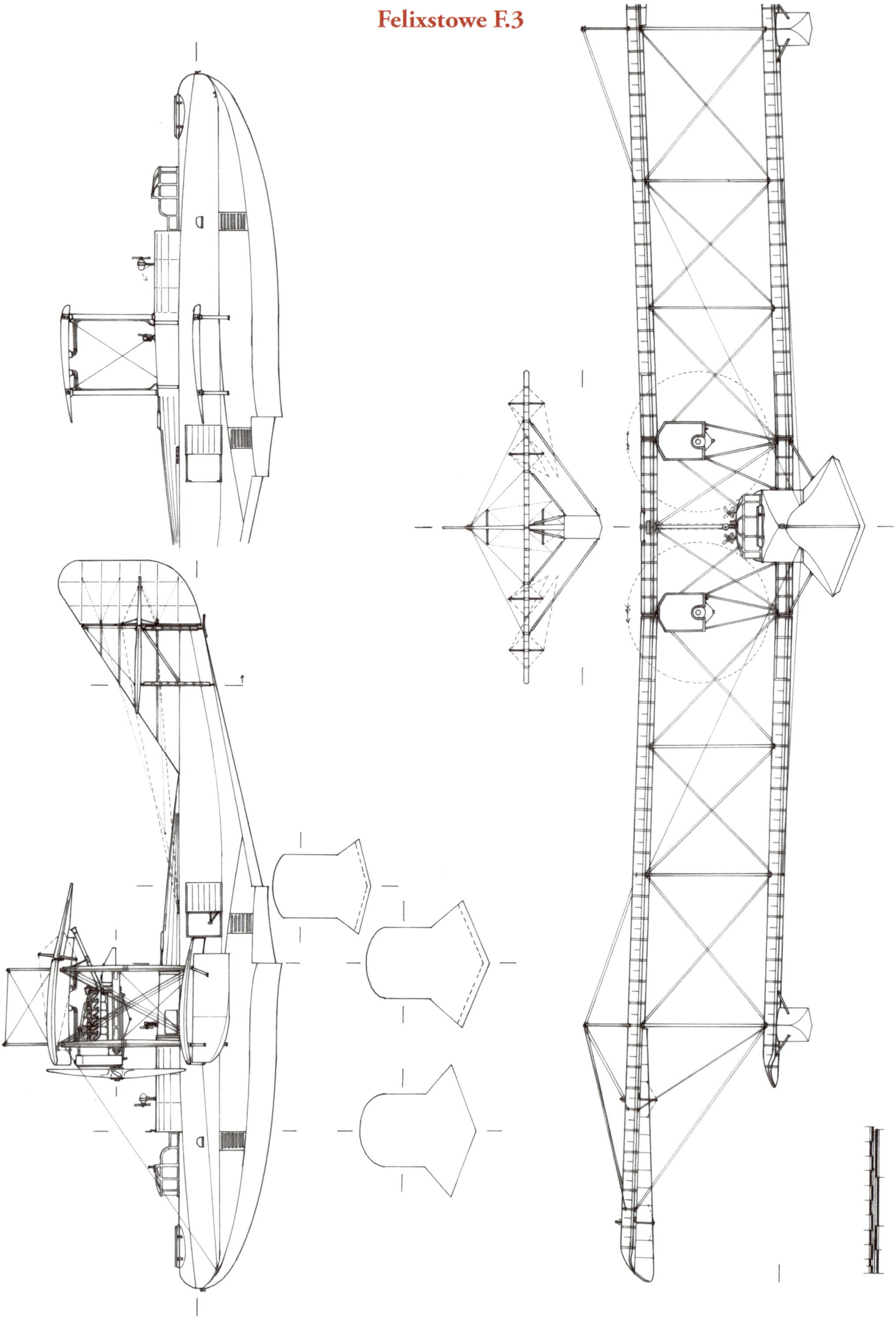

Felixstowe F.3

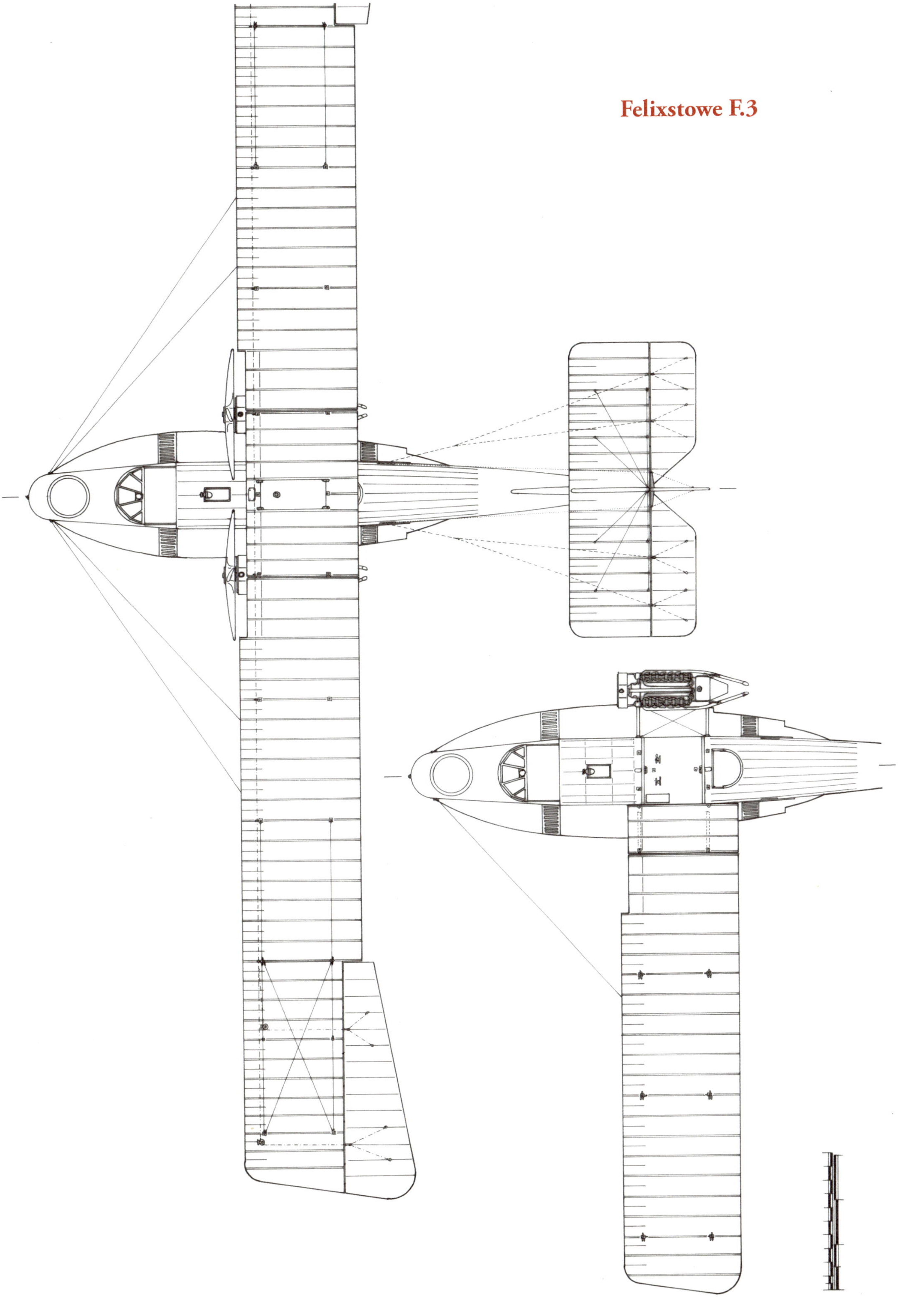

Felixstowe F.3

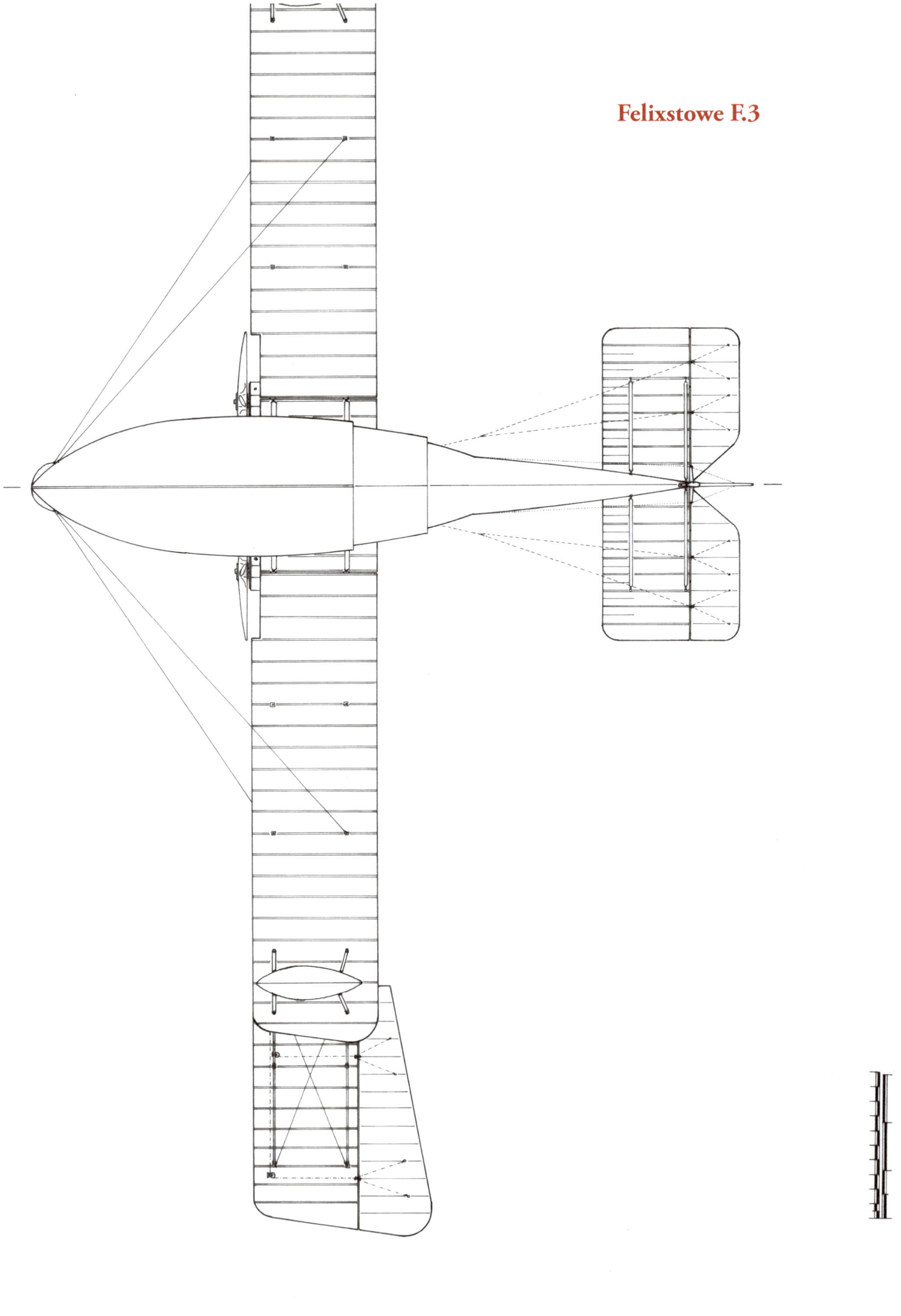

Made in the USA
Las Vegas, NV
10 November 2026

57846470R00164